T0213528

The Next Era in Hardware Security

Nikhil Rangarajan • Satwik Patnaik
Johann Knechtel • Shaloo Rakheja
Ozgur Sinanoglu

The Next Era in Hardware Security

A Perspective on Emerging Technologies for Secure Electronics

 Springer

Nikhil Rangarajan
Division of Engineering
New York University Abu Dhabi
Abu Dhabi
United Arab Emirates

Satwik Patnaik
Department of Electrical and Computer
Engineering
Texas A&M University
College Station
TX, USA

Johann Knechtel
Division of Engineering
New York University Abu Dhabi
Abu Dhabi
United Arab Emirates

Shaloo Rakheja
Holonyak Micro and Nanotechnology
Laboratory
University of Illinois at Urbana Champaign
Urbana
IL, USA

Ozgur Sinanoglu
Division of Engineering
New York University Abu Dhabi
Abu Dhabi
United Arab Emirates

ISBN 978-3-030-85794-3   ISBN 978-3-030-85792-9   (eBook)
https://doi.org/10.1007/978-3-030-85792-9

This Springer imprint is published by the registered company Springer Nature Switzerland AG
The registered company address is: Gewerbestrasse 11, 6330 Cham, Switzerland

We acknowledge all the efforts by the medical community, the research community, and everyone else affected by and dealing with the COVID-19 pandemic. We also extend our condolences to those who have lost loved ones.—Nikhil, Satwik, Johann, Shaloo, and Ozgur

To Anja and Richard. I am truly grateful for having you both in my life. Richard, may you have a bright future; be curious, be courageous, and be humble on all your paths through life.—Johann

To my wife and my parents, who have been my pillar of strength and whose constant support has helped me through times thick and thin.—Nikhil

Foreword

With technology scaling reaching nanometer range and the ever-increasing demands for new application domains, emerging technologies, different from conventional CMOS VLSI, are seriously being investigated to complement it. Emerging technologies based on novel materials and new computing paradigms represent an alternate paradigm in the quest for secure and robust electronics. Conventional logic and memory systems have traditionally employed security as an afterthought, if at all, at the circuit or system design stage. However, emerging technologies and novel materials can offer a unique opportunity to incorporate security-centric design practices at the device conception stage itself. By promoting device-circuit co-design with security as a prime metric (along with the traditional power, performance, and area considerations), emerging technologies may indeed reinvent the way we think about hardware security.

This book provides well-curated insights for the potential of emerging devices, materials, and the resulting novel architectures to tackle various challenges of hardware security, ranging from intellectual property protection to data security. Each chapter describes a particular set of properties, which are promising to advance the state of the art for the implementation of security schemes. Broad and informative reviews of seminal emerging technologies and their related security schemes are included, along with detailed case studies to highlight the most important aspects of emerging secure electronics. This book aims to educate the reader on the various approaches and practices employed for forging emerging technologies into viable security schemes. The tutorial-style writing and organization aids the reader through the whole process, starting from (1) a pedagogical "connecting the dots" approach for mapping the needs and requirements of particular security primitives with the properties offered by emerging technologies, over to (2) reviews of relevant concepts, and (3) detailed case studies, which can inspire readers to apply their own knowledge of emerging technologies and security, to not only learn but possibly devise their own schemes.

Karlsruhe, Germany   Mehdi Tahoori
June 2021
                                                                                              M. Tahoori

Preface

This book describes recent research at the confluence of hardware security and emerging technologies and complements it thoroughly with fundamentals of both domains. This book focuses on the unique properties of emerging technologies pertinent to the needs of hardware security and discusses methodologies to leverage those properties in order to build secure computing and storage systems. It presents these security-centric device characteristics in a structured and logical manner, along with recent research results, which serve well for both educational purposes and to motivate further interest and research in this field.

More specifically, this book is a comprehensive compilation of hardware security concepts borne out of the unique characteristics of emerging logic and memory devices, and related architectures. The primary focus is on mapping emerging technology-specific properties like multi-functionality, runtime polymorphism, intrinsic entropy, nonlinearity, ease of heterogeneous integration, tamper-resilience, and side-channel resistance to the corresponding security primitives that they help realize, namely static and dynamic camouflaging, true random number generation, physical unclonable functions, secure heterogeneous and large-scale systems, tamper-proof hardware, and side-channel-resilient circuits, respectively. We discuss several technologies offering the desired properties, including but not limited to spintronics switches, memristors, silicon nanowire transistors, and optoelectronic devices, for such security primitives and schemes, while also providing detailed case studies for each of the outlined security application. Overall, the book aims to provide a holistic perspective of how the promising properties found in emerging technologies, which are not readily afforded by traditional CMOS devices and systems, can help advance the field of hardware security.

This book is applicable and relevant for graduate students, researchers, and practitioners in the field of electrical and computer engineering with particular interest in hardware security and emerging technologies. Given the structured approach of the book, with good emphasis on the foundations for different emerging technologies and their promising properties (which are later on explored in detail for various case studies on novel security schemes), this book would serve well for readers seeking to broaden their scope, both as a reference for foundational concepts and state of the art in the field and as an inspiration for their own future work.

Abu Dhabi, United Arab Emirates   Nikhil Rangarajan

College Station, TX, USA   Satwik Patnaik

Abu Dhabi, United Arab Emirates   Johann Knechtel

Urbana-Champaign, IL, USA   Shaloo Rakheja

Abu Dhabi, United Arab Emirates   Ozgur Sinanoglu
June 2021

Acknowledgements

The case studies in this monograph are based on selected research publications of the authors. The authors would like to acknowledge the contributions of Ramesh Karri, Mohammed Nabeel, Mohammed Ashraf, Arun Parthasarathy, Yogesh S. Chauhan, Jörg Henkel, Hussam Amrouch, Jacek Gosciniak, Mahmoud Rasras, Alabi Bojesomo, and the late Vassos Soteriou, who served as co-authors in those respective publications.

Contents

Acronyms

ADC	Analog to Digital Converter
AES	Advance Encryption Standard
AFM	Antiferromagnet
AFMRAM	Antiferromagnetic Random Access Memory
AHB	Advanced High Performance Bus
AHE	Anomalous Hall Effect
AMBA	Advanced Microcontroller Bus Architecture
APU	Address Protection Unit
ASIC	Application Specific Integrated Circuit
ASL	All-Spin Logic
ATPG	Automatic Test Pattern Generation
BEOL	Back-End-Of-Line
BI	Bus Interface
BUF	Buffer
CAD	Computer-Aided Design
CCR	Correct Connection Rate
CEP	Common Evaluation Platform
CLB	Configurable Logic Block
CME	Counter Mode Encryption
CMOS	Complementary Metal Oxide Semiconductor
CNOT	Controlled-NOT
CNT	Carbon Nanotube
CNTFET	Carbon Nanotube Field Effect Transistor
CPA	Correlation Power Analysis
CTS	Clock-Tree Synthesis
DAC	Digital to Analog Converter
DFT	Discrete Fourier Transform
DIMM	Dual In-line Memory Module
DIP	Distinguishing Input Pattern
DoS	Denial of Service
DPA	Differential Power Analysis

DPU	Data Protection Unit
DRAM	Dynamic Random Access Memory
DRC	Design Check Rule
DW	Domain Wall
DWM	Domain Wall Motion
ECC	Error Correction Code
ECP	Error Correction Pointer
EDA	Electronic Design Automation
EM	Electromagnetic
F2F	Face-to-Face
FC	Fault Coverage
FDTD	Finite-Difference Time-Domain
FE	Ferroelectric
FeFET	Ferroelectric Field Effect Transistor
FEOL	Front-End-Of-Line
FFT	Fast Fourier Transform
FHD	Fractional Hamming Distance
FIB	Focused Ion Beam
FM	Ferromagnet
FPGA	Field Programmable Gate Array
GPS	Global Positioning System
GSHE	Giant Spin-Hall Effect
HD	Hamming Distance
HPC	High Performance Computing
HR	High Resistance
HT	Hardware Trojan
HWSF	Hardware Security Feature
I/O	Input/Output
IC	Integrated Circuit
INV	Inverter
IP	Intellectual Property
ISEA	Interposer-based Security-Enforcing Architecture
LC	Layout Camouflaging
LL	Logic Locking
LR	Low Resistance
LSPR	Localized Surface Plasmon Resonance
LUT	Look-Up Table
M3D	Monolithic 3D
MAC	Message Authentication Code
MAJ	Majority
ME	Magnetoelectric
MESO	Magnetoelectric Spin-Orbit
MI	Magnetic Insulator
MITM	Man-In-The-Middle
MOSFET	Metal Oxide Semiconductor Field Effect Transistor

MTJ	Magnetic Tunnel Junction
MUX	Multiplexer
NBTI	Negative Bias Temperature Instability
NCFET	Negative Capacitance Field Effect Transistor
NM	Nanomagnet
NoC	Network on Chip
NP	Nanoparticle
NVM	Non-Volatile Memory
OER	Output Error Rate
OTP	One Time Pad
P/G	Power/Ground
PCC	Pearson Correlation Coefficient
PCM	Phase Change Memory
PDN	Power Distribution Network
PNR	Percentage of Netlist Recovery
PoC	Proof of Concept
PPA	Power Performance Area
PRS	Policy Register Space
PSC	Power Side Channel
PUF	Physically Unclonable Function
PVC	Passive Voltage Contrast
PVT	Process Voltage Temperature
RAP	Resistance Area Product
RDL	Redistribution Layer
RE	Reverse Engineering
RFID	Radio Frequency Identification
RISC	Reduced Instruction Set Computer
RLL	Random Logic Locking
RO	Ring Oscillator
RoT	Root of Trust
RRAM	Resistive Random Access Memory
RTL	Register Transfer Level
SAF	Slave Access Filter
SAT	Boolean Satisfiability
SCA	Side-Channel Analysis
SDR	Silicon Disk Resonator
SEM	Scanning Electron Microscope
SFLL	Stripped Functionality Logic Locking
SiNWFET	Silicon Nanowire Field Effect Transistor
s-LLGS	stochastic Landau Lifshitz Gilbert Slonczewski
SMART	Secure Magnetoelectric Antiferromagnetic Tamper-Proof Memory
SoC	System on Chip
SPA	Simple Power Analysis
SPEF	Standard Parasitic Exchange Format
SPP	Surface Plasmon Polariton

SRAM	Static Random Access Memory
SRS	Shared Register Space
STT	Spin-Transfer Torque
STT-MRAM	Spin-Transfer Torque Magnetoresistive Random Access Memory
SVF	Side-channel Vulnerability Factor
TCL	Tool Command Language
TCU	Trusted Configuration Unit
TI	Topological Insulator
TIGFET	Three-Independent-Gate Field Effect Transistor
TMD	Transition Metal Dichalcogenide
TMO	Transition Metal Oxide
TMR	Tunneling Magnetoresistance
TRANSMON	Transaction Monitor
TRNG	True Random Number Generator
TSC	Thermal Side Channel
TSV	Through-Silicon Via
TVC	Threshold Voltage-based Camouflaging
VCD	Value Change Dump
VLSI	Very Large-Scale Integration
VTOPSS	Voltage-controlled Topological Spin Switch
WDDL	Wave Dynamic Differential Logic
WLCSP	Wafer-Level Chip-Scale Packaging

Chapter 1
Introduction

1.1 Fundamentals of Hardware Security

In our modern age of omnipresent and highly interconnected information technology, cybersecurity becomes ever more challenged. For example, with the rise of the Internet of Things (IoT), most such equipment is connected to the internet in some way, often inscrutable to the regular customers. This fact opens up large attack surfaces and can lead to severe ramifications (Fig. 1.1), as demonstrated through a plethora of real-world attacks over many years. Within the realm of cybersecurity, hardware security in particular is concerned about achieving security and trust directly within the underlying electronics. For example, researchers have cautioned against powerful attacks on the speculative execution of modern processors [Koc+19a, Lip+18a] or profiled the side-channel leakage of cryptographic hardware modules [Ler+18]. For another example, the so-called root of trust (RoT) techniques for isolation and attestation of computation are found in many commercial computers and other custom devices [Mae+18, Zha+19, Nab+20].

Next, we discuss the fundamental aspects of hardware security and review selected prior art, which will be assumed as common knowledge throughout the remaining chapters of this monograph. It should be understood that this chapter can provide only an overview on this vast and fast-growing field, but we review the most important aspects and seminal protection schemes here to equip the reader with the necessary background to follow this monograph.

1.1.1 Data Security at Runtime

The confidentiality, integrity, and availability of data processing within electronics are subject to various threat scenarios, like (1) unauthorized access or modification

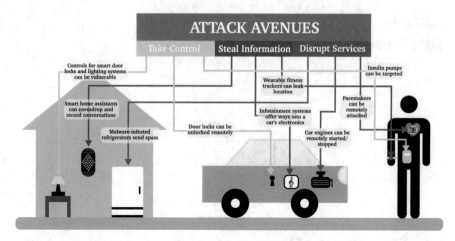

Fig. 1.1 Modern devices are deployed all around us and often connected to the internet but typically lack built-in notions and measures for security. Thus, a plethora of risks arise for everyday life. Adopted from [Gra16]

of data and (2) attacks leveraging side-channels, fault-injection, physical read-out or probing.

1.1.1.1   Unauthorized Access or Modification of Data

Conventional attacks seeking to steal or corrupt data are conducted mainly at the software level and for interconnected systems. Cryptography represents a commonly applied protection scheme here, but there are also many dedicated, hardware-centric security features. For example, there are:

- Enclaves for trusted execution (TEEs), like the industrial *ARM TrustZone* and *Intel SGX* or the academic *MIT Sanctum* (Fig. 1.2; see also [Mae+18] for more background on each TEE)
- Wrappers for monitoring and cross-checking of untrusted third-party intellectual property (IP) modules [Bas+17]
- Centralized IP infrastructures for secure system design [Wan+15b]
- Verification of computation [Wah+16]
- Secure task scheduling [Liu+14]
- Secure network-on-chip (NoC) architectures [Fio+08], etc.

However, if not designed and implemented carefully, such security features become prone to hardware-centric attacks themselves, e.g., see [BCO04, Bay+16, Qiu+19, OD19, CH17]; such attacks are discussed in more detail throughout this section.

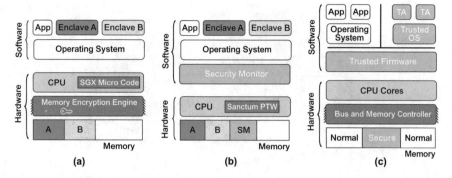

Fig. 1.2 Enclaves for trusted execution are prominent hardware-centric security features. Illustrated are the high-level architectures of (**a**) *Intel SGX*, (**b**) *MIT Sanctum*, and (**c**) *ARM TrustZone*

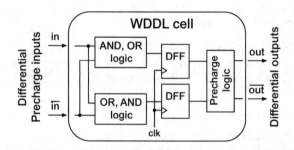

Fig. 1.3 Wave dynamic differential logic (WDDL) serves to mitigate power side-channel attacks. This is achieved by redundant, differential combinational logic paths that switch in opposite directions for any operation, thereby obfuscating differences in power consumption for particular transitions. Furthermore, pre-charge logic is used to reduce peak power consumption during switching. Adopted from [Fuj+14]

1.1.1.2 Side-Channel and Fault-Injection Attacks

Side-channel attacks infer information from physical channels that are leaky due to the sensitivity and vulnerability of the underlying electronics [ZF05]. For example, it is well-known that the Advanced Encryption Standard (AES) is vulnerable to power side-channel attacks when the hardware implementation is unprotected [BCO04, OD19, SW12]. Another instance is the leakage of information related to timing behavior or speculative execution in modern processors, through caches and other buffers [OST05, Lip+18a, Sch+19].

Most countermeasures against side-channel attacks apply some kind of cloaking or masking technique, which involves the diffusion of the information leaked through side-channels. This is achieved by various means, ranging from system-level solutions [GMK16], down to individual gates [Bel+18]. See also Fig. 1.3 for an example. Nevertheless, the resilience of such countermeasures is still subject to the physical implementation in its entirety. For example, Fujimoto et al. [Fuj+14]

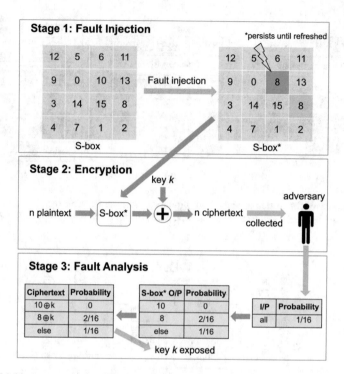

Fig. 1.4 Multi-stage attack on AES. First, a random but persistent fault is injected into the S-box, to bias the subsequent encryption. In turn, the bias helps to infer the secret key while sampling over a sufficiently large number of cipher-texts. Adopted from [Pan+19]

have shown that even the promising WDDL scheme suffers from minute layout-level asymmetries, allowing an attacker to eventually infer the secret key.

Fault-injection attacks induce faults to deduce sensitive information. Therefore, fault injection can also support or advance other attacks; see Fig. 1.4 for an example. Fault-injection attacks cover (1) direct, invasive fault injection, e.g., by laser light [SHS16] or electromagnetic waves [CH17, Bay+16, Deh+12], as well as (2) indirect fault injection, e.g., by repetitive writing to particular memory locations [Vee+16] or by deliberate "misuse" of dynamic voltage and frequency scaling (DVFS) features [Qiu+19].

Countermeasures include detection of faults at runtime [Nat+19] and hardening against fault injection at design and manufacturing time [Kar+18b, LM06, Dut+18]. Note that distinguishing between natural and malicious faults is non-trivial [Kar+18a], which imposes practical challenges for recovery at runtime.

1.1.1.3 Physical Read-Out and Probing Attacks

An adversary with access to equipment used traditionally for failure analysis or inspection, like electro-optical probing or focused ion beam milling tools [Pri+17a], can mount quite powerful read-out attacks. Among others, these attacks include:

1. Probing of transistors and wires [Wan+17a, Hel+13], either through the metal layers or the substrate backside
2. Monitoring the photon emission induced by CMOS transistor switching [Taj+17, Kra+21]
3. Monitoring the electrical charges in memories [CSW16]

When applied carefully, these attacks can reveal *all* internal signals. For instance, Fig. 1.5 shows the concept and an example of an electro-optical probing technique, allowing to infer data/bits of individual devices at runtime.

Countermeasures seek to prevent and/or detect the physical access. Prior solutions have sought to place shielding structures in the back-end-of-line (BEOL) [Wan+19, Lee+19, YPK16], deflection or scrambling structures in the substrate [She+18] (see also Fig. 1.6), and detector circuitry [Wei+18]. Earlier studies such as [ISW03] also considered formally secure techniques. However, such schemes are subject to limitations assumed for the attackers, which can become obsolete and would then render the formal guarantees void.

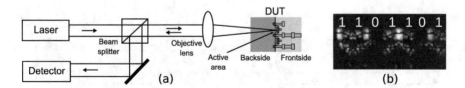

Fig. 1.5 Concept (**a**) and example (**b**) for laser voltage probing, which serves to read-out data/bits in devices at runtime. In the example (**b**), an array of registers is in view, with those remaining dark storing "0" and those lighting up storing "1." Subfigure (**a**) is adopted from [Loh+16] and (**b**) adopted from a related microscopy image provided as courtesy by Shahin Tajik

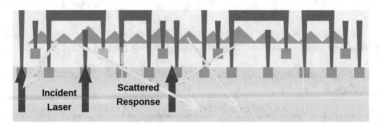

Fig. 1.6 Concept of pyramid structures in silicon substrate, e.g., achieved using dedicated etching steps. As a result, reflections from incident laser light are scattered and intermingled, making a distinct read-out of data more difficult. Adopted from [She+18]

1.1.2 Securing the Integrity and Confidentiality of Hardware

Besides the severe threats on data security at runtime, as outlined above, other threats such as reverse engineering (RE), piracy of chip-design intellectual property (IP), illegal overproduction, counterfeiting, or insertion of hardware Trojans represent further challenges for hardware security. These threats arise due to the globalized and distributed nature of modern supply chains for electronics, which span across many entities and countries. See also Fig. 1.7 for an overview of related threats.

A multitude of protection schemes have been proposed, which can be broadly classified into IP protection, Trojan defense, and physically unclonable functions (PUFs). IP protection can be further broadly classified into logic locking, camouflaging, and split manufacturing. All these schemes seek to protect the hardware from different attack scenarios, which include untrusted foundries, untrusted testing facilities, untrusted end-users, or a combination thereof. For example, selected techniques for IP protection are illustrated in Table 1.1 along with the related untrusted entities.

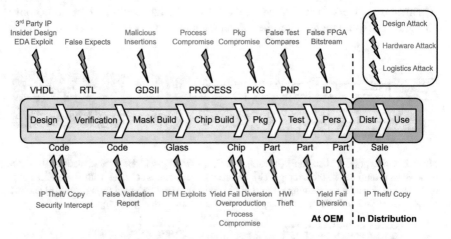

Fig. 1.7 An overview on design-, hardware-, and logistics-centric attacks throughout the (largely outsourced) supply chain of modern electronics. Adopted from [Ber16]

Table 1.1 IP protection techniques versus untrusted entities (✓: Protection offered, ✗: No protection offered)

Technique	FEOL/BEOL foundry	Test facility	End-user
Logic locking	✓/✓	✓ (see also [Yas+16a])	✓
Layout camouflaging	✗/✗ (✓/✗ [Pat+17, Pat+20a], ✓/✓ [Ran+20a])	✗ (✓ [Ran+20a])	✓
Split manufacturing	✓/✗ (✗/✓ [Wan+17c])	✗	✗ (✓ [Pat+19a])

1.1.2.1 Logic Locking

Logic locking protects the design IP by inserting dedicated key-gates, which are operated by a secret key [YRS20]. Without the knowledge of the secret key, logic locking ensures that the details of the design IP cannot be fully recovered and the IC remains non-functional. The key-gates are commonly realized using, e.g., XOR/XNOR gates [RKM10], AND/OR gates [Dup+14], or look-up tables (LUTs) [BTZ10]. Only after manufacturing (preferably even after testing [Yas+16a]) is the IC to be activated, by loading the secret key into a dedicated, on-chip tamper-proof memory. The realization of such tamper-proof memories remains a topic of active research and will be covered in more detail in Chap. 7.

Threat Model In general, a threat model describes the capabilities of the attacker and the resources at their disposal. It also classifies entities as trusted or untrusted. The threat model for logic locking can be summarized as follows:

- The design house, designers, and the electronic design automation (EDA) tools the designers work with are considered *trusted*, whereas the foundry, the test facility, and the end-user(s) are all considered *untrusted*.
- The attackers possess knowledge regarding the logic locking technique that has been applied to protect the design IP.
- The attackers have access to the locked netlist (e.g., by RE). Hence, they can identify the key inputs and the related logic but are oblivious to the secret key.
- The secret key cannot be tampered with, as it is programmed in a tamper-proof memory.
- The attackers are in possession of a functional chip bought from the open market. This chip can act as an "oracle" for evaluating input/output patterns.

Without knowledge of the secret key, logic locking ensures that: (1) the details of the original design cannot be fully recovered; (2) the IC is non-functional, i.e., it produces incorrect outputs.

Logic Locking Techniques and Attacks Early research proposed random logic locking (*RLL*) [RKM10], fault-analysis-based locking (*FLL*) [Raj+15], and strong interference-based locking (*SLL*) [Raj+12], all to protect against brute force attacks. These techniques identify suitable, selected locations for inserting the key-gates. However, multiple attacks have undermined the security guarantees of these aforementioned techniques by formulating different attacks [Raj+12, PM15, LO19].

In 2015, Subramanyan et al. [SRM15] challenged the security promises of all then-known logic locking techniques. This attack leveraged Boolean satisfiability (SAT) to compute the so-called *discriminating input patterns (DIPs)*. A DIP generates different outputs for the same input across two (or more) different keys, indicating that at least one of the keys is incorrect. The attack stepwise evaluates different DIPs until all incorrect keys have been pruned. The attack experiences its worst case scenario when it can eliminate only one incorrect key per DIP; here, $2^k - 1$ DIPs are required to resolve k key bits. In general, the SAT attack

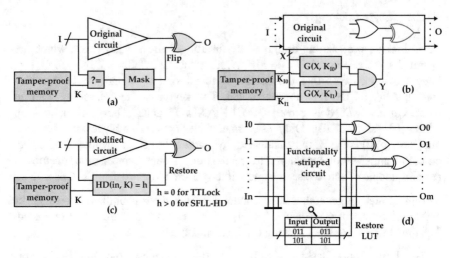

Fig. 1.8 Selected SAT-resilient locking techniques: (**a**) *SARLock*, (**b**) *Anti-SAT*, (**c**) *TTLock/SFLL-HD*, and (**d**) *SFLL-flex*. Adopted from [YS17]

resilience of any locking technique can be represented by the number of DIPs required to decipher the correct key and the average time taken for each SAT attack iteration [XS16].

Initial research in SAT-resilient logic locking techniques aimed to increase the complexity of the SAT-based attack by ensuring that the attack pruned out *exactly* one incorrect key per iteration. To that end, in 2016, *SARLock* [Yas+16b] and *Anti-SAT* [XS16] were put forward as defense techniques against the SAT-based attack [SRM15]. *SARLock* (Fig. 1.8a) employs a controlled corruption of the output, across all incorrect keys, for exactly one input pattern. *SARLock* can also be integrated with other high-corruptibility techniques (e.g., *FLL* or *SLL*) to provide a two-layer defense. In *Anti-SAT* [XS16], two complementary logic blocks, embedded with the key-gates, converge at an AND gate (Fig. 1.8b). The output of this AND gate is always "0" for the correct key; for the incorrect key, it may be "1" or "0," depending on the inputs. This AND gate then feeds an additional XOR gate that is interposed into the original design, thereby possibly inducing incorrect outputs for incorrect keys. Both techniques utilize the concept of *point functions* and enforce low output corruptibility to obtain resilience against the SAT-based attack.

The two-layer defense of *SARLock* was *approximately* circumvented by *App-SAT* [Sha+18a] and *Double DIP* [SZ17]. In both the attacks, the combination of a low-corruption part (resilient to SAT attacks) and a high-corruption part (prone to SAT attacks) is reduced to the low-corruption part (e.g., *SARLock* + *SLL* to *SARLock*). The *Double DIP* [SZ17] can eliminate at least two incorrect keys in each iteration, thereby increasing the attack efficiency. For *Anti-SAT*, the two complementary blocks at its heart exhibit significant signal skews, rendering them distinguishable from other logic, which is exploited by Yasin et al. in the *signal probability skew (SPS)* attack [Yas+16c]. Moreover, both *SARLock* and *Anti-SAT* are vulnerable to the *bypass attack* [Xu+17]. This attack picks some key randomly and determines the inputs that provide incorrect outputs for this chosen key. Then, additional logic is constructed around the *Anti-SAT/SARLock* blocks to recover the overall circuit from these incorrect outputs.

In *TTLock*, the original logic is modified for exactly one input pattern [Yas+17a]. The output for this protected pattern is restored using a comparator block, as illustrated in Fig. 1.8c. Even if an attacker succeeds to remove the comparator block, she/he obtains a design different from the original one (albeit for only one input pattern). On the heels of *TTLock*, Yasin et al. [Yas+17b] proposed *stripped functionality logic locking (SFLL)*. SFLL is resilient against most current attacks, and it enables one to trade-off between resilience against SAT attacks and removal attacks [Yas+17b]. It is based on the notion of "strip and restore," where some part of the original design is removed and the intended functionality is concealed.

SFLL has three variants, *SFLL-HD* [Yas+17b], *SFLL-flex* [Yas+17b], and *SFLL-rem* [Sen+20], which we briefly discuss next. *SFLL-HD* is a generalized version of *TTLock* that allows the designer to protect a larger number of selected input patterns. It should be noted that *SFLL-HD* protects a restricted set of input cubes, which are all underpinned by one secret key. *SFLL-flex$^{c \times k}$*, in contrast, allows to protect any c selected input cubes, each with k specified bits. Here, the protected patterns are typically represented using a small set of input cubes, which are then stored in an on-chip LUT (Fig. 1.8d). In *SFLL-fault*, fault-injection-based heuristics are leveraged to identify and protect multiple patterns and to reduce area cost at the same time. Both *SFLL-HD* and *SFLL-flex$^{c \times k}$* utilize AND-trees that leave structural hints for an opportune attacker. Related attacks have been demonstrated by Sirone et al. [SS19] and Yang et al. [YTS19], tackling these locking schemes without access to an oracle. Recently, graph neural network-based unlocking techniques [Alr+20] have been proposed to facilitate this notion of oracle-less recovery attacks as well. The sparse prime implicant (SPI) attack [HYR21] was shown to break the security guarantees of *SFLL-rem* [Sen+20].

Further Defenses and Attacks Related to Logic Locking Shamsi et al. [Sha+18b] presented a layout-centric logic locking scheme, based on routing cross-bars comprising obfuscated and configurable vias. The notion of *cyclic locking* has been proposed in [Sha+17b] and extended in [RMS18]. The idea is to create supposedly unresolvable locking instances by introducing feedback cycles. However, tailored SAT formulations have challenged such locking schemes [ZJK17, She+19].

Recent works have proposed *parametric locking* [Yas+17a, XS17, Zam+18, CXS18]; the essence is to lock design parameters and profiles. For example, in [XS17], the key not only protects the functionality of the design but also its timing profile. A functionally correct but timing-incorrect key will result in timing violations, thereby leading to circuit malfunctions. A *timing-based SAT attack*, presented in [CLS18], circumvented the timing locking approach in [XS17]. Therefore, further research into *parametric locking* is required. The notion of *mixed-signal locking* has been advocated recently as well, e.g., in [Jay+18, Leo+19].

Finally, further attack/defense schemes on regular locking techniques have been proposed that focus on inferring/obfuscating the structural modifications induced by locking, without requiring an oracle, e.g., [CCB18, Alr+21].

1.1.2.2 Layout Camouflaging

Layout camouflaging serves to mitigate RE attacks conducted by malicious end-users. Broadly speaking, layout camouflaging alters the layout-level appearance of an IC in order to protect the design IP. As illustrated in Fig. 1.9, layout camouflaging can be achieved by dedicated front-end-of-line (FEOL) processing steps, like manipulation of dopant regions, gate structures, and/or gate contacts [Raj+13a, Erb+16, Li+16], but also by obfuscation of the back-end-of-line (BEOL) interconnects [Pat+17, Pat+20a]. Layout camouflaging has been made available for commercial application, e.g., see the *SypherMedia Library* [Ram19]. Note that obfuscation is also known in the context of design-time protection, e.g., by obfuscating finite state machines [LP15]—such techniques are orthogonal to layout camouflaging.

Threat Model The threat model for layout camouflaging is summarized as follows:

- The design house and foundry are trusted, the test facility is either trusted or untrusted, and the end-user is untrusted.
- The adversary holds one or multiple functional chip copies and is armed with sophisticated equipment and know-how to conduct RE. The resilience of any camouflaging scheme ultimately depends on the latter.
- The adversary is aware of the camouflaging scheme, and she/he can identify the camouflaged gates, infer all the possible functions implemented by the camouflaged cell, but cannot readily infer the actual functionality.

Layout Camouflaging Techniques and Attacks Similar to logic locking, early studies focused on the selection of gates to camouflage (and the design of camouflaged cells). In their seminal work, Rajendran et al. [Raj+13a] proposed a camouflaged NAND-NOR-XOR cell. The authors also proposed clique-based

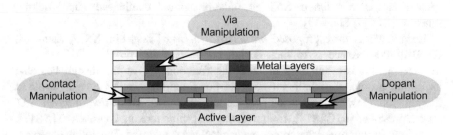

Fig. 1.9 Device-level concepts for layout camouflaging

selection for camouflaging, based on their own finding that a random selection of gates to camouflage can be resolved by sensitization-based attacks [Raj+13a].

Massad et al. [MGT15] and Yu et al. [Yu+17] independently formulated SAT-based attacks (with oracle access), which challenged the security guarantees of [Raj+13a]. These attacks could readily circumvent small-scale camouflaging for various benchmarks with up to 256 gates being camouflaged. A parallel SAT-based attack providing an average speedup of 3.6× over prior attacks was presented by Wang et al. [Wan+18a]. Keshavarz et al. [Kes+18] proposed a SAT-based formulation augmented by probing and fault-injection capabilities, where the authors were able to RE an *S-Box*. Still, it remains to be seen whether the attack can tackle larger designs.

In [YSR17], Yasin et al. demonstrated how an untrusted test facility can circumvent the security promise of camouflaging, even without access to an oracle. The authors deciphered the layout camouflaging technique presented in [Raj+13a] successfully by analyzing the test patterns provided by the design house to the untrusted test facility. To the best of our knowledge, none of the layout camouflaging techniques proposed thus far have been able to mitigate this kind of attack, except for the *dynamic camouflaging* technique presented in [Ran+20a], which is discussed in more detail in Chap. 3.

Many existing layout camouflaging techniques, e.g., [Raj+13a, Nir+16, CMG16, Wan+16a], exhibit a significant cost with respect to power, performance, and area (PPA). For example in [Raj+13a], camouflaging 50% of the design results in ≈150% overheads for power and area, respectively. See Fig. 1.10 for an analytical experiment on layout costs for camouflaging versus split manufacturing (introduced in the next subsection). A more comprehensive investigation of PPA cost for various layout camouflaging techniques is provided in [Pat+17, Pat+20a].

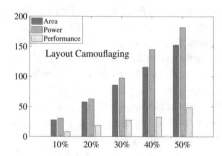 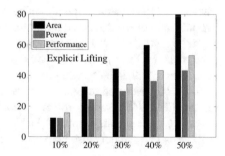

Fig. 1.10 Study on PPA cost (%) for layout camouflaging [Raj+13a] (left) and lifting of wires (randomly selected) to M8 (metal layer 8) for split manufacturing (right). Results are averaged across *ITC-99* benchmarks. For layout camouflaging (left), the impact on power and area is substantial, given that the NAND–NOR–XOR structure proposed in [Raj+13a] incurs 4× and 5.5× more area and power compared to a regular 2-input NAND gate. For split manufacturing (right), the cost for area is severe; that is because routing resources are relatively scarce for M8 (pitch = 0.84 μm) and lifting of wires occupies further routing resources, which can only be obtained by enlarging the die outlines

Most layout camouflaging techniques also require modifications in the front-end-of-line (FEOL) manufacturing process, which can incur financial cost on top of PPA overheads. Therefore, in such scenarios, layout camouflaging is applied selectively, to limit PPA cost and the impact on FEOL processing. However, the selective application of layout camouflaging techniques compromises the security guarantees, especially in the light of *oracle-guided SAT* attacks such as [MGT15, Yu+17, Wan+18a].

The notion of *provably secure layout camouflaging* was put forward in [Yas+16d, Li+16]. *CamoPerturb* [Yas+16d] seeks to minimally perturb the functionality of the design by either removing or adding one *minterm* (i.e., the product term of all variables). A separate block, called *CamoFix*, is then added to restore the minterm; *CamoFix* is built up using camouflaged INV/BUF cells. This concept is inherently similar to the idea of TTLock, which was discussed previously. Inspired by logic locking, Li et al. [Li+16] employ AND-trees as well as OR-trees for layout camouflaging. Depending on the desired security level, tree structures inherently present in the design are leveraged, or additional trees are inserted. Then, the inputs of the trees are camouflaged using dopant-obfuscated cells.

Both techniques [Yas+16d, Li+16] have been shown to exhibit vulnerabilities: [Li+16] was circumvented by *sensitization-guided SAT attack (SGS)* [Yas+17c], while Jiang et al. [Jia+18] circumvented *CamoPerturb* using *sensitization* and *implication* principles leveraged from automated test pattern generation (ATPG). In general, these techniques are also vulnerable to *approximate* attacks outlined in [Sha+17a, Sha+18a]. A follow-up work to [Li+16] is presented in [Li+17], where the authors discuss how structural attacks like *SPS* [Yas+16c] can be rendered ineffective when the trees are obfuscated both structurally and functionally. However, such structural and functional obfuscations are also vulnerable to sophisticated attacks.

Further Defenses and Attacks Related to Layout Camouflaging Besides the various analytical attacks, RE may also compromise layout camouflaging techniques directly. For example, ambiguous gates [Raj+13a, Coc+14] or secretly configured MUXes [Wan+16a] rely on dummy contacts and/or dummy channels, which will induce different charge accumulations at runtime. Courbon et al. [CSW16] leveraged scanning electron microscopy in the passive voltage contrast mode (SEM PVC) for measurement of charge accumulations, where they succeeded in reading out a secured memory. Furthermore, monitoring the photon emission at runtime can presumably also help uncover layout camouflaging [Loh+16].

Threshold voltage-based layout camouflaging (TVC) has gained traction over the past few years. The essence of TVC is a selective manipulation of dopants at the transistor level that creates logic cells that are identical structurally but depict different functionalities. Nirmala et al. [Nir+16] proposed TVC cells that can operate as NAND, NOR, OR, AND, XOR, or XNOR. Erbagci et al. [Erb+16] proposed TVC cells operating as XOR or XNOR, based on the selective use of high- and low-threshold transistors. Collantes et al. [CMG16] adopted *domino logic* to implement their TVCs. Recently, Iyengar et al. [Iye+18] demonstrated two flavors of TVC in STMicroelectronics 65 nm technology. In principle, TVC techniques offer better resilience than other layout camouflaging techniques, as regular etching and optical imaging techniques are ineffective with respect to TVC. Still, TVC may be revealed eventually, e.g., by leveraging SEM PVC [Sug+15].

Another interesting avenue is the camouflaging of the back-end-of-line (BEOL), i.e., the interconnects [Che+15, Pat+17, Pat+20a, Jan+18]. Chen et al. [Che+15, CCW18] explored the use of real vias (magnesium, Mg) along with dummy vias (magnesium oxide, MgO) to achieve the same. They have shown that Mg can oxidize quickly into MgO, thereby hindering an identification by an RE attacker. Recently, Patnaik et al. [Pat+17, Pat+20a] extended the concept of BEOL camouflaging in conjunction with split manufacturing, to protect against an untrusted FEOL foundry. Patnaik et al. developed customized cells and design stages for BEOL camouflaging, where they successfully demonstrated full-chip camouflaging at lower PPA cost than prior works. Their study also explored how large-scale (BEOL) camouflaging can thwart SAT-based attacks, by inducing overly large and complex SAT instances, which overwhelm the SAT solvers.

Most techniques discussed so far cannot be configured post-fabrication, i.e., they implement static camouflaging. In contrast, Akkaya et al. [AEM18] demonstrated a reconfigurable camouflaging scheme that leverages hot-carrier injection. Notably, the authors fabricated a prototype in 65 nm technology; however, they report significant PPA cost (e.g., in comparison to regular NAND gates, they report $9.2\times, 6.6\times$, and $7.3\times$ for power, performance, and area, respectively). Zhang et al. [Zha+18a] introduced the concept of *timing-based camouflaging*, based on wave-pipelining and false paths. However, this technique was circumvented in [Li+18b].

1.1.2.3 Split Manufacturing

Split manufacturing seeks to protect the design IP from untrustworthy foundries [RSK13, Sen+17, Pat+18e, Pat+18f, McC16, Pat+21]. As indicated by the term, the idea is to split the manufacturing flow, most commonly into an untrusted FEOL process and a subsequent, trusted BEOL process (Fig. 1.11).

Such splitting into FEOL and BEOL is practical for multiple reasons:

- Outsourcing the FEOL is desired, as it requires some high-end and costly facilities.
- BEOL fabrication on top of the FEOL is significantly less complex than FEOL fabrication itself.
- Some in-house or trusted third-party facility can be engaged for BEOL fabrication.
- The sole difference for the supply chain is the preparation and shipping of FEOL wafers to that facility for BEOL fabrication.

In fact, split manufacturing has been demonstrated successfully; [Vai+14a] describes promising results for a 130 nm process split between *IBM* and *Global-*

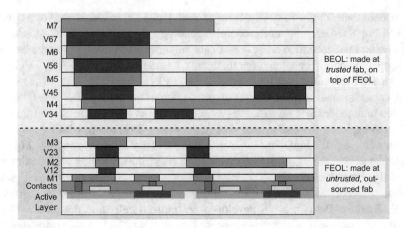

Fig. 1.11 Classical split manufacturing, i.e., the separation into front-end-of-line (FEOL) and back-end-of-line (BEOL) parts

Foundries, and [McC16] reports on a 28 nm split process run by *Samsung* across Austin and South Korea.

Regarding the security notion of split manufacturing, a split layout appears to an adversary in the FEOL facility as a "sea of incomplete gates and wires," making it challenging to infer the entire netlist, its design IP, and ultimately the functionality. Still, given that regular, security-agnostic design tools work holistically on both the FEOL and BEOL, hints on the missing wiring can remain in the FEOL [Wan+16b, Li+19], which can be exploited by attackers, to decipher the missing connections.

Threat Model The basic, most common threat model for split manufacturing is summarized as follows:

- The design house and end-user are trusted, while the FEOL foundry is deemed untrustworthy. Split manufacturing necessitates a trusted BEOL foundry, with assembly and testing facilities typically also considered as trustworthy. The end-user is also considered trustworthy (e.g., defense establishments).
- With the design house and end-user being trusted, the adversary cannot obtain a chip copy from those entities. Besides, the chip has typically not been manufactured before; the chip is then unavailable altogether for RE attacks.
- The primary goal of the adversary is to infer the missing BEOL connections from the incomplete FEOL layout. Once the attacker deciphers all the missing connections correctly, he/she can pirate and/or illegally overproduce the design IP. To that end, she/he (1) is aware of the underlying protection technique, if any, and (2) has access to the EDA tools, libraries, and other supporting information.

An "inverted threat model" was explored in [Wan+17c], where the BEOL facility is untrustworthy and the FEOL fab is trustworthy. Since fabricating the

FEOL is more costly than the BEOL, the practical relevance of this model remains questionable.

Another variation of the threat model was explored recently by Chen and Vemuri [CV18b]. The authors assume that a working chip is available, which is then used as an oracle for a SAT-based formulation to recover the missing BEOL connections. While it is not explicitly stated in [CV18b], we presume that the authors seek to recover the gate-level details of some design whose functionality is otherwise already available/known. For an attacker, doing so can be relevant, e.g., for inserting hardware Trojans during re-implementation of some existing design, or to obtain the IP without RE of the available chip copy.

Imeson et al. [Ime+13a] further proposed a strong model in the context of hardware Trojans. Here, the attacker already holds the design netlist and is interested in inserting Trojans in appropriate locations. This work [Ime+13a], also known as *k-security*, has been further extended in [Li+18a].

Split Manufacturing Techniques and Attacks The first attack on split manufacturing was proposed by Rajendran et al. [RSK13]. The notion of this so-called *proximity attack* is as follows. Although the layout is split into FEOL and BEOL, it is still designed holistically (when using regular EDA tools). Rajendran et al. [RSK13] infer the missing connections in the BEOL from the proximity of cells, which is readily observable in the FEOL. While this attack shows a good accuracy for small designs, the same is not true for larger designs. Wang et al. [Wan+16b] extended this attack, by taking into account a multitude of FEOL-level hints: (1) physical proximity of gates, (2) avoidance of combinatorial loops (which are rare in practice for combinational designs), (3) timing and load constraints, and (4) directionality of "dangling wires" (i.e., the wires remaining unconnected in the top-most FEOL layer). Magaña et al. [MSD16] proposed various routing-based attack techniques and conclude that such attacks are more effective than solely placement-centric attacks. Recently, Zhang et al. [ZMD18] and Li et al. [Li+19] leveraged machine learning techniques for deciphering missing connections. However, neither attack [MSD16, ZMD18, Li+19] recovers the actual netlist; rather, they provide the most probable BEOL connections.

On the defense side, various techniques have been proposed to safeguard FEOL layouts against proximity attacks, e.g., [RSK13, Vai+14b, Wan+16b, Sen+17, Wan+17d, MSD16, Fen+17, Pat+18e, Pat+18f, CV18a, Li+20, Pat+21]. They can be categorized into (1) placement-centric, (2) routing-centric, and (3) both placement- and routing-centric defenses.

Wang et al. [Wan+16b] and Sengupta et al. [Sen+17] propose placement perturbation. Layout randomization is the most secure technique, especially when splitting after the first metal layer, as shown by Sengupta et al. [Sen+17]. However, this technique has limited scalability and demonstrates significant layout cost for larger designs. In general, placement-centric works caution that splitting after higher metal layers—which helps to limit financial cost and practical hurdles for split manufacturing [XFT15, Pat+18e, Pat+18f, Pat+21]—can undermine their resilience. That is because any placement perturbation is eventually offset by routing at higher layers.

Routing-centric techniques such as those in [RSK13, Wan+17d, MSD16, Fen+17, Pat+18e, Li+20, Pat+21] resolve proximity and other hints at the FEOL routing. Rajendran et al. [RSK13] proposed to swap pins of IP modules and to re-route those nets, thereby obfuscating the design hierarchy. As these swaps cover only part of the interconnects, this technique cannot protect against gate-level IP piracy. In fact, 87% of the connections could be correctly recovered on the *ISCAS-85* benchmarks [RSK13]. In general, routing-centric techniques are subject to the available routing resources and PPA budgets, which can ease proximity attacks. For example, [Wan+17d, Fen+17] consider short routing detours, and [MSD16] consider few routing blockages to limit impact on design timing.

Patnaik et al. [Pat+18e, Pat+21] proposed various heuristics as well as custom cells for lifting wires to the BEOL in a concerted manner. The authors demonstrated a superior resilience against the state-of-the-art network-flow attack [Wan+16b] and deep learning attacks [Li+19] when compared to the prior placement- and routing-centric techniques. Later on, Patnaik et al. [Pat+18f] proposed randomization at the netlist level, which is carried through the EDA flow, thereby resulting in an erroneous and misleading FEOL layout. The original design is only restored at the BEOL, using customized routing cells. This work is one of the first to address holistic protection of both placement and routing, which also demonstrated superior protection against the state-of-the-art proximity attacks.

Further Defenses and Attacks Related to Split Manufacturing Inspired by logic locking, Sengupta et al. [Sen+19] realize IP protection at manufacturing time by locking the FEOL and subsequent unlocking of the BEOL. The authors also formalize the problem of split manufacturing, borrowing concepts from logic locking.

As mentioned before, Imeson et al. [Ime+13a] formulated the notion of k-*security* to prevent targeted insertion of hardware Trojans. The idea is to create k isomorphic structures in the FEOL by guided lifting of wires to the BEOL. Now, an attacker cannot uniquely map these k structures to some specific target in the already known design; she/he has to either randomly guess (with a probability of $1/k$) or insert multiple Trojans. Li et al. [Li+18a] extended k-*security* in various ways. Most notably, they leverage additional gates and wires to elevate the security level beyond those achieved in [Ime+13a]. Recently, Xu et al. [Xu+19] questioned the theoretical security of k-*security* by pattern matching attacks.

Vaidyanathan et al. [VDP14] advocate testing of the untrusted FEOL against Trojan insertion, using BEOL stacks dedicated for testability. Xiao et al. [XFT15] propose the notion of obfuscated built-in self-authentication (OBISA) to hinder IP piracy and Trojan insertion.

While advanced attacks such as [ZMD18, Li+19, Li+20] are on the rise, split manufacturing becomes inherently more resilient for larger, industrial designs. In fact, none of the existing attacks have succeeded in completely recovering all missing BEOL connections for larger designs yet. Still, the premise for split manufacturing—to resolve hints from the FEOL—remains. Thus, schemes that further reduce the dependency on EDA tools (and cost) are required. Although [Sen+19] explore the formalism of split manufacturing, the notion of provably secure split manufacturing remains an open problem.

Finally, "entering the next dimension of split manufacturing," by leveraging the up-and-coming techniques for 3D integration, has been initiated in [Val+13, Ime+13a, Kne+17, Pat+18c, Gu+18b, Pat+19b]. Further research toward this end seems promising as well.

1.1.2.4 Trojan Defense

The notion of Trojans is wide-ranging and requires multiple dimensions for classification [BT18]—it relates to malicious hardware modifications that are (1) working at the system level, register-transfer level (RTL), gate/transistor level, or the physical level; (2) leveraging the digital and/or the physical domain; (3) seeking to leak information from an IC, reduce the IC's performance, or disrupt the working of the IC altogether; (4) are always on, triggered internally, or triggered externally. For example for (2), digital Trojans are activated by either a specific, rare input pattern or via "time bombs" on certain operations (or input patterns) being executed for a particular number of cycles. On the other hand, physical Trojans are activated either by detrimental effects such as electromigration, negative bias temperature instability, etc., or by internal or external side-channel triggers.

Trojans are likely introduced by untrustworthy third-party IP, adversarial designers, or through "hacking" of computer-aided design (CAD) tools [Bas+19], or, arguably even more likely, during distribution and deployment of ICs [Swi+17]. Although it has been projected traditionally as the main scenario, we argue that the likelihood of Trojans being introduced at fabrication time is rather low. That is because any such endeavor, once detected, would fatally disrupt the business of the affected foundry. Therefore, foundries can be expected to employ technical and organizational means available to them to hinder modifications by malicious employees.

Defense techniques can be classified into (1) Trojan detection during design and manufacturing time and (2) Trojan mitigation at runtime. The former relies on testing and verification steps [Cha+09a, Aar+10, JM08, LJM12, Guo+19, Sug+15, Vas+18, Cha+15], whereas the latter relies on dedicated security features for testability and self-authentication [XFT14], current monitoring [GBF17], monitoring and detection of malicious activities [KV11, Bhu+13, Bas+17, Wu+16, Wah+16], etc. See also Fig. 1.12 for an example of the latter features. Note that the two classes (Trojan detection during design and manufacturing time for one, Trojan mitigation at runtime for another) may also intersect, for example with the use of built-in self-authentication modules [Shi+17].

Besides, IP protection schemes like logic locking and split manufacturing can hinder Trojan insertion at manufacturing time, at least to a certain degree. That is

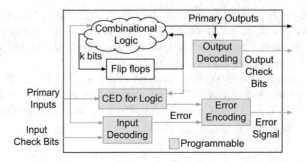

Fig. 1.12 Trojan mitigation at runtime, based on concurrent error detection (CED) and input/output as well as error encoding modules. The latter are required for the overall system, where multiple chips of the outlined architecture will be linked together for cross-verification. Adopted from [Wu+16]

because an adversary without the full understanding of the layout and its IP cannot easily insert specific, targeted Trojans [Ime+13a, Pat+19b].

1.1.2.5   Physically Unclonable Functions

When applied some input stimulus, a PUF should provide a fully de-correlated output response. This response must be reproducible for the very same PUF, even under varying environmental conditions, but it must differ across different PUF instances, even for the same PUF design. The desired properties for PUFs are uniqueness, unclonability, unpredictability, reproducibility, and tamper-resilience.

PUFs are used for (1) "fingerprinting" or authentication of hardware, using the so-called "weak PUFs" that provide capabilities for processing only one/few fixed inputs, or (2) challenge-response-based security schemes, using the so-called "strong PUFs" that provide capabilities for processing a large number of inputs [Her+14, MV10, CZZ17]. Note that "weak PUFs" are not necessarily inferior to "strong" PUFs [Rüh+13a]. On the contrary, powerful machine learning attacks such as [CZZ17, Rüh+13a, Liu+18, Gan17] do *not* apply for weak PUFs, only for strong PUFs. The main difference between weak and strong PUFs is that, as indicated, the former work on a (few) fixed input(s), or inputs or *challenge(s)*, whereas the latter have to support a large range of challenges. See Fig. 1.13 for the outline of a default, generic authentication scheme using a strong PUF.

Electronic PUFs represent the dominant class of PUFs, with prominent types of electronic PUFs using ring oscillators, arbiters, bistable rings, and memories [MV10, Her+14, Gan17, CZZ17]. Such PUFs are relatively simple to implement and integrate, even for advanced processing nodes. The core principle for

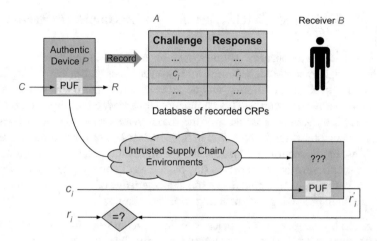

Fig. 1.13 A default, generic authentication scheme using a strong PUF. The trusted entity A collects a set L of challenge-response pairs (C, R) generated from the PUF P and stores L in a database for interrogation in the field. P is then delivered to B so that the PUF can be used by B for authentication with A. To do so, A randomly selects an unseen challenge c_i, sends it to B, who applies it to P to generate the response r_i'. This response is sent back, whereupon A checks r_i' against the corresponding response r_i in its database; in case these responses match (subject to error tolerances as allowed for by the particular protocol), B is successfully authenticated using P. Adopted from [CZZ17]

such PUFs is to leverage the process variations inherent to CMOS fabrication and operation, through various dedicated circuitry. However, the resulting randomness is limited for most CMOS PUF implementations; it may be machine-learned and, thus, cloned [CZZ17, Rüh+13a, Liu+18, Gan17].

Optical PUFs represent another interesting class [Pap+02, Rüh+13b, TŠ07, MV10, Gru+17, Kne+19]. In fact, the very first PUF proposal in the literature, proposed by Pappu et al. [Pap+02] in 2002, devised an optical token from transparent epoxy with randomly inserted, micrometer-sized glass spheres. Thus, the idea of optical PUFs is to manufacture an "optical token" that, in addition to structural variations inherently present in selected optical media, may contain randomly included materials, e.g., nanoparticles. Besides such a token, optical PUFs require further components, for generating the optical input and processing the output. The fundamental phenomena underlying an optical PUF are scattering, reflection, coupling, and absorption of light within the optical token. Depending on the materials used for the token and the inclusions as well as the design of the token itself, these phenomena can be highly chaotic and nonlinear by nature [Kne+19, Gru+17]. Hence, optical PUFs are considered more powerful than electronic PUFs.

1.2   Limitations of CMOS Technology for Hardware Security

Most of the hardware security primitives we have visited in the previous sections have been predominantly CMOS-centric. However, emerging devices bring a new facet to this equation, by offering unique properties that can reshape the way we think about hardware security. This confluence of hardware security and emerging devices has been gaining traction over the recent years owing to new physics and materials research, which has resulted in the development of novel logic and memory devices. In this section, we first discuss the limitations of CMOS technology for hardware security, which have driven this push toward emerging technology-based solutions, before delving into the specific characteristics of emerging technologies that make them promising for upcoming security primitives:

1. In general, emerging technologies seek to overcome the fundamental CMOS limitations regarding power consumption, among other aspects. Power overheads are a crucial consideration while securing any CMOS IP, with full-chip protection often requiring a large sacrifice in terms of constraining the power budget. This typically prohibits the percentage of the chip that we can feasibly secure. On the other hand, many emerging devices have been shown to exhibit ultra-low operating power, thus enabling the designer to secure a larger portion of the chip without exacerbating power consumption.
2. Similarly, the scalability of CMOS-based security primitives has been a cause for concern, since they can often result in blowing up the chip area, thus increasing the fabrication costs. In contrast, emerging devices cannot only possess a smaller device footprint, but in most cases also allow the designer to implement logic more efficiently, thus saving valuable die area.
3. The existing CMOS framework has always been built with the primary intent of improving the performance and efficiency of modern electronics, with security being retrofitted as an afterthought. Emerging devices afford us the opportunity to embed security as a primary design metric in the supply chain and allow a security-focused device-circuit co-design process.
4. Conventional CMOS logic styles and synthesis techniques are well documented and have been around for a long time. These do not offer much flexibility in terms of patching circuit- and system-level security vulnerabilities. However, emerging devices can potentially allow completely new logic design styles to counter those vulnerabilities, owing to their unique construction and operation. For instance, novel ferroelectric and spintronic logic devices with non-volatility can enable memory-in-logic, thus opening the doors to new non-von Neumann computing paradigms.

It must be noted that although these emerging technologies appear promising and practical for the near future, they will likely be built to augment the existing CMOS framework, not supplant or replace it. In this context hybrid CMOS-emerging electronics such as the N3XT architecture [Aly+18], which combine carbon nanotubes and spintronics within CMOS 3D ICs, might not be a distant reality.

1.3 Inherent Properties of Emerging Technologies to Advance Hardware Security

As mentioned in Sect. 1.2, various emerging technologies offer the potential to advance the notion of hardware security. Figure 1.14 outlines selected emerging technologies, their properties relevant and beneficial for hardware security, the security schemes that are supported accordingly, and the security threats countered by such schemes.

The emerging devices included in Fig. 1.14 have some interesting properties in common, which are more difficult to achieve in traditional CMOS technology. More specifically, spintronic devices, memristors, carbon nanotube field effect transistors (CNTFETs), and Silicon Nanowire field effect transistors (SiNWFETs) can all be tailored to achieve significant variability/randomness, reconfigurability or polymorphic behavior, resilience against reverse engineering, heterogeneous integration, and also the possibility of separating trusted and untrusted parts. Therefore, these devices can serve well for PUFs, TRNGs, IP protection schemes, and to mask side-channel leakage. Moreover, memristors may also offer resilience against tampering, by means of destructive data management. It should be understood that the prospects for actual implementation of such security schemes based on emerging devices depend on various aspects, ranging from circuit design and security analysis in general, down to manufacturing capabilities and device maturity, among others. Next, we briefly discuss some of these unique properties of emerging devices, which make them prime candidates for the next era of secure electronics.

1.3.1 Reconfigurability

Reconfigurability, in the context of emerging devices, refers to the ability of a single device topology to be configured for different logic functionalities, depending on external control signals or internal parameters like doping, etc. Essentially, it means that the same device layout could possibly implement one of several logic gates, each indistinguishable from each other. However, once configured and deployed in the field, the reconfigurable device will retain its functionality. Yet, the very fact that the particular functionality it implements could be one of many, increases the computational complexity for an attacker seeking to decipher its true nature. This is particularly useful for static camouflaging schemes, where the number of functionalities that the emerging logic gate can implement decides the number of key bits per camouflaged gate. Here, the multi-functional gate could be configured

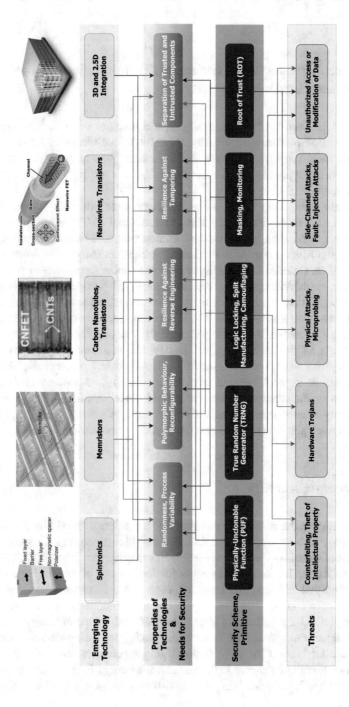

Fig. 1.14 A selective overview on emerging technologies, their properties, matching security schemes, and countered threats

for its intended logic, post-fabrication, in a secure facility. Some classes of devices that exhibit reconfigurability include the giant spin Hall effect device [Ran+19], SiNWFETs [Rai+18], and memristors [Xia+09].

1.3.2   Runtime Polymorphism

The ability to undergo reconfiguration in the field is referred to as runtime polymorphism in this monograph. Runtime polymorphic emerging devices cannot only be programmed for functionality once post-fabrication, but can be morphed any number of times on-the-fly. This characteristic makes them useful for implementing polymorphic logic for hardware security. The difference between reconfigurable and polymorphic logic from the context of the attacker is that while reconfigurable gates present the attacker with the conundrum of correctly identifying the true nature of the logic from many possible functionalities, polymorphic logic actually morphs between those various possible functional states in real-time. Thus, an attacker seeking to RE the polymorphic circuit not only has to identify these functional states but also the transformation scheme between the states. Polymorphic emerging devices proposed in prior works include the magnetoelectric spin–orbit device (MESO) [Man+19] and magnetic domain wall (DW)-based devices [Par+17].

1.3.3   Nonlinearity and Intrinsic Entropy

Nonlinearity in an emerging device refers to its ability to perform nonlinear transformation operations on the input stimulus, wherein the input to output mapping becomes nonlinear or deviates from direct proportionality. Such an injective one-way nonlinear transformation is quite useful for generating keys in a physical unclonable function setting. Examples of nonlinear emerging devices finding applications in cryptographic key generation include memristors [Zha+18b] and photonic micro-cavity-based devices [Gru+17].

Intrinsic entropy of an emerging device is the innate randomness in the physical phenomenon driving the device. Such randomness could arise due to any number of factors such as metastability or abruptness in switching processes, or chaotic dynamics. Evidence of intrinsic entropy has been observed in several emerging devices such as spintronics switches [Ven+15] and photonic systems [SR07].

1.3.4   Heterogeneous Physical Integration

Emerging technologies like 3D and 2.5D integration allow for the bifurcation of the chip into security-critical and non-critical components, thus enabling schemes like split manufacturing to come into the picture. The possibility of physical separation and integration of the chip in the supply chain enables the designer to incorporate safeguards against IP theft and insertion of malicious Trojans. Besides, the heterogeneous integration of novel emerging devices into hybrid-CMOS configurations within the 3D framework opens up opportunities to tap the unique characteristics of those devices, for e.g., in-memory computing in 3D architecture [Aly+18].

1.3.5   Resilience Against Tampering and Side-Channel Attacks

Tamper- and side-channel resilience in an emerging device or technology arises from the ability to shield its internal operation and switching activity from an outsider. This could be achieved via a number of techniques like materials engineering to make the device impervious to external stimuli from probes, or constructing physical barriers to obstruct an attacker from accessing internal nodes and wires. Immunity against side-channel attacks also stems from features that aid in securing the leakage channels, like masking or modifying the power activity visible to the attacker. For instance, triple-independent gate field effect transistors (TIGFET) [Sha+20] and all-spin logic [AYL18] have been shown to be resistant to power side-channel attacks due to symmetric I-V and symmetric read/write characteristics, respectively.

1.4   Closing Remarks

Our future electronics and computing systems can be fortified against various classes of hardware threats by looking beyond conventional CMOS design practices, toward the vast array of emerging logic and memory technologies. Some of these emerging devices have shown promise in mitigating the common hardware-based concerns known to plague CMOS systems, owing to their unique and peculiar properties. In this chapter, we first looked at the fundamentals of hardware security, including the various threat scenarios and attack landscapes against modern computing systems, and the plethora of solutions proposed to counter them. Then we tried to understand the limitations of CMOS hardware with respect to securing hardware, at the root level, and also delved into the specific properties of emerging technologies that can address these limitations. These properties, viz. reconfigurability, runtime polymorphism, nonlinearity, intrinsic entropy, heterogeneous physical integration, tamper-resilience, and resistance against side-channel leakage, form the basis of individual security primitives that will be discussed in the upcoming chapters.

References

[Aar+10] J. Aarestad et al., Detecting trojans through leakage current analysis using multiple supply pad IDDQS. Trans. Inf. Forens. Sec. **5**(4), 893–904 (2010). https://doi.org/10.1109/TIFS.2010.2061228

[AEM18] N.E.C. Akkaya, B. Erbagci, K. Ma, A secure camouflaged logic family using postmanufacturing programming with a 3.6GHz adder prototype in 65nm CMOS at 1V nominal VDD, in *Proc. Int. Sol.-St. Circ. Conf.* (2018)

[Alr+20] L. Alrahis et al., GNNUnlock: graph neural networks-based oracle-less unlocking scheme for provably secure logic locking. Preprint . arXiv:2012.05948 (2020)

[Alr+21] L. Alrahis et al., UNSAIL: Thwarting oracle-less machine learning attacks on logic locking. Trans. Inf. Forens. Sec. **16**, 2508–2523 (2021)

[Aly+18] M.M.S. Aly et al., The N3XT approach to energy-efficient abundant-data computing. Proc. IEEE **107**(1), 19–48 (2018)

[AYL18] Q. Alasad, J. Yuan, J. Lin, Resilient AES against side-channel attack using all-spin logic, in *Proceedings of the 2018 on Great Lakes Symposium on VLSI* (2018), pp. 57–62

[Bas+17] A. Basak et al., Security assurance for system-on-chip designs with untrusted IPs. Trans. Inf. Forens. Sec. **12**(7), 1515–1528 (2017). ISBN: 1556-6013. https://doi.org/10.1109/TIFS.2017.2658544

[Bas+19] K. Basu et al., CAD-base: an attack vector into the electronics supply chain. Trans. Des. Autom. Elec. Sys. **24**(4), 38:1–38:30 (2019). ISBN: 1084-4309. https://doi.org/10.1145/3315574

[Bay+16] P. Bayon et al., Fault model of electromagnetic attacks targeting ring oscillator-based true random number generators. J. Cryptogr. Eng. **6**(1), 61–74 (2016). ISSN: 2190-8516. https://doi.org/10.1007/s13389-015-0113-2

[BCO04] E. Brier, C. Clavier, F. Olivier, Correlation power analysis with a leakage model, in *Proc. Cryptogr. Hardw. Embed. Sys.* (2004)

[Bel+18] D. Bellizia et al., Secure double rate registers as an RTL countermeasure against power analysis attacks. Trans. VLSI Syst. **26**(7), 1368–1376 (2018). ISSN: 1063-8210. https://doi.org/10.1109/TVLSI.2018.2816914

[Ber16] K. Bernstein, Rapid authentication through verification, validation, and marking. Tech. rep. 2016. https://www.ndia.org/-/media/sites/ndia/meetings-and-events/divisions/trusted-micro-electronics--joint-working-group/687c-aug-2016/2-bernstein---distro-a.ashx

[Bhu+13] S. Bhunia et al., Protection against hardware trojan attacks: towards a comprehensive solution. Des. Test **30**(3), 6–17 (2013). ISSN: 2168-2356. https://doi.org/10.1109/MDT.2012.2196252

[BT18] S. Bhunia, M.M. Tehranipoor (eds.) *The Hardware Trojan War: Attacks, Myths, and Defenses* (Springer, Berlin, 2018). https://doi.org/10.1007/978-3-319-68511-3

[BTZ10] A. Baumgarten, A. Tyagi, J. Zambreno, Preventing IC piracy using reconfigurable logic barriers. Des. Test **27**(1), 66–75 (2010). ISSN: 0740-7475

[CCB18] P. Chakraborty, J. Cruz, S. Bhunia, SAIL: machine learning guided structural analysis attack on hardware obfuscation, in *Proc. Asian Hardw.-Orient. Sec. Trust Symp.* (2018), pp. 56–61. https://doi.org/10.1109/AsianHOST.2018.8607163

[CCW18] S. Chen, J. Chen, L. Wang, A chip-level anti-reverse engineering technique. J. Emerg. Tech. Comput. Syst. **14**(2), 29:1–29:20 (2018). ISSN: 1550-4832. https://doi.org/10.1145/3173462

[CH17] A. Cui, R. Housley, BADFET: defeating modern secure boot using second-order pulsed electromagnetic fault injection, in *Proc. Worksh. Off. Tech.* (2017). https://www.usenix.org/conference/woot17/workshop-program/presentation/cui

[Cha+09a] R.S. Chakraborty et al., MERO: a statistical approach for hardware trojan detection, in *Proc. Cryptogr. Hardw. Embed. Sys.* (2009), pp. 396–410. https://doi.org/10.1007/978-3-642-04138-9_28

[Cha+15] A. Chandrasekharan et al., Ensuring safety and reliability of IP-based system design – A container approach, in *Proc. Int. Symp. Rapid System Prototyping* (2015), pp. 76–82. https://doi.org/10.1109/RSP.2015.7416550

[Che+15] S. Chen et al., Chip-level anti-reverse engineering using transformable interconnects, in *Proc. Int. Symp. Def. Fault Tol. in VLSI Nanotech. Sys.* (2015), pp. 109–114. https://doi.org/10.1109/DFT.2015.7315145

[CLS18] A. Chakraborty, Y. Liu, A. Srivastava, TimingSAT: timing profile embedded SAT attack, in *Proc. Int. Conf. Comp.-Aided Des.* (2018), pp. 6:1–6:6

[CMG16] M.I.M. Collantes, M.E. Massad, S. Garg, Threshold-dependent camouflaged cells to secure circuits against reverse engineering attacks, in *Proc. Comp. Soc. Symp. VLSI* (2016), pp. 443–448. https://doi.org/10.1109/ISVLSI.2016.89

[Coc+14] R.P. Cocchi et al., Circuit camouflage integration for hardware IP protection, in *Proc. Des. Autom. Conf.* (2014), pp. 1–5. https://doi.org/10.1145/2593069.2602554

[CSW16] F. Courbon, S. Skorobogatov, C. Woods, Direct charge measurement in floating gate transistors of flash EEPROM using scanning electron microscopy, in *Proc. Int. Symp. Test. Failure Analys.* (2016), pp. 1–9. https://pdfs.semanticscholar.org/992a/20c0a8bb71642fc44fa65f053b3524113b99.pdf

[CV18a] S. Chen, R. Vemuri, Improving the security of split manufacturing using a novel BEOL signal selection method, in *Proc. Great Lakes Symp. VLSI* (2018), pp. 135–140

[CV18b] S. Chen, R. Vemuri, On the effectiveness of the satisfiability attack on split manufactured circuits, in *Proc. VLSI SoC* (2018)

[CXS18] A. Chakraborty, Y. Xie, A. Srivastava, GPU obfuscation: attack and defense strategies, in *Proc. Des. Autom. Conf.* (2018), pp. 122:1–122:6

[CZZ17] C.H. Chang, Y. Zheng, L. Zhang, A retrospective and a look forward: fifteen years of physical unclonable function advancement. IEEE Circ. Syst. Mag. **17**(3), 32–62 (2017). ISSN: 1531-636X. https://doi.org/10.1109/MCAS.2017.2713305

[Deh+12] A. Dehbaoui et al., Electromagnetic transient faults injection on a hardware and a software implementations of AES, in *Proc. Worksh. Fault Diag. Tol. Cryptogr.* (2012), pp. 7–15. https://doi.org/10.1109/FDTC.2012.15

[Dup+14] S. Dupuis et al., A novel hardware logic encryption technique for thwarting illegal overproduction and hardware trojans, in *Proc. Int. On-Line Test Symp.* (2014), pp. 49–54

[Dut+18] J.-M. Dutertre et al., Laser fault injection at the CMOS 28 nm technology node: an analysis of the fault model, in *Proc. Worksh. Fault Diag. Tol. Cryptogr.* (2018). https://doi.org/10.1109/FDTC.2018.00009

[Erb+16] B. Erbagci et al., A secure camouflaged threshold voltage defined logic family, in *Proc. Int. Symp. Hardw.-Orient. Sec. Trust* (2016), pp. 229–235. https://doi.org/10.1109/HST.2016.7495587

[Fen+17] L. Feng et al., Making split fabrication synergistically secure and manufacturable, in *Proc. Int. Conf. Comp.-Aided Des.* (2017)

[Fio+08] L. Fiorin et al., Secure memory accesses on networks-on-chip. Trans. Comput. **57**(9), 1216–1229 (2008). ISSN: 0018-9340. https://doi.org/10.1109/TC.2008.69

[Fuj+14] D. Fujimoto et al., Correlation power analysis using bit-level biased activity plaintexts against AES cores with counter-measures, in *International Symposium on Electromagnetic Compatibility* (2014)

[Gan17] F. Ganji, On the learnability of physically unclonable functions. PhD thesis. Technische Universität Berlin, 2017. https://doi.org/10.14279/depositonce-6174

[GBF17] L.A. Guimarães, R.P. Bastos, L. Fesquet, Detection of layout-level trojans by monitoring substrate with preexisting built-in sensors, in *Proc. Comp. Soc. Symp. VLSI* (2017), pp. 290–295. https://doi.org/10.1109/ISVLSI.2017.58

[GMK16] H. Groß, S. Mangard, T. Korak, Domain-oriented masking: compact masked hardware implementations with arbitrary protection order, in *Proc. Comp. Comm. Sec.* (2016). https://doi.org/10.1145/2996366.2996426

[Gra16] A. Grau, How to build a safer internet of things: todays IoT is full of security flaws. We must do better. IEEE Spectr. (2016). http://spectrum.%20ieee.%20org/telecom/security/how-tobuild-a-safer-internet-of-things

[Gru+17] B.C. Grubel et al., Silicon photonic physical unclonable function. Opt. Express **25**(11), 12710–12721 (2017). https://doi.org/10.1364/OE.25.012710

[Gu+18b] P. Gu et al., Cost-efficient 3D integration to hinder reverse engineering during and after manufacturing, in *Proc. Asian Hardw.-Orient. Sec. Trust Symp.* (2018), pp. 74–79

[Guo+19] X. Guo et al., QIF-Verilog: quantitative information-flow based hardware description languages for pre-silicon security assessment, in *Proc. Int. Symp. Hardw.-Orient. Sec. Trust* (2019), pp. 91–100. https://doi.org/10.1109/HST.2019.8740840

[Hel+13] C. Helfmeier et al., Breaking and entering through the silicon, in *Proc. Comp. Comm. Sec.* Berlin, Germany (2013), pp. 733–744. ISBN: 978-1-4503-2477-9. https://doi.org/10.1145/2508859.2516717

[Her+14] C. Herder et al., Physical unclonable functions and applications: a tutorial. Proc. IEEE **102**(8), 1126–1141 (2014). ISSN: 0018-9219. https://doi.org/10.1109/JPROC.2014.2320516

[HYR21] Z. Han, M. Yasin, J. (JV) Rajendran, Does logic locking work with EDA tools?, in *Proc. USENIX Sec. Symp.* (2021). https://www.usenix.org/conference/usenixsecurity21/presentation/han-zhaokun

[Ime+13a] F. Imeson et al., Securing computer hardware using 3D integrated circuit (IC) technology and split manufacturing for obfuscation, in *Proc. USENIX Sec. Symp.* (2013)

[ISW03] Y. Ishai, A. Sahai, D. Wagner, Private circuits: securing hardware against probing attacks, in *Advances in Cryptology* (2003), pp. 463–481. ISBN: 978-3-540-45146-4. https://doi.org/10.1007/978-3-540-45146-4_27

[Iye+18] A.S. Iyengar et al., Threshold defined camouflaged gates in 65nm technology for reverse engineering protection, in *Proc. Int. Symp. Low Power Elec. Design* (2018), pp. 6:1–6:6

[Jan+18] J.-W. Jang et al., Threshold-defined logic and interconnect for protection against reverse engineering. Trans. Comput. Aided Des. Integr. Circ. Syst. **39**, 308 (2018)

[Jay+18] N.G. Jayasankaran et al., Towards provably-secure analog and mixed-signal locking against over-production, in *Proc. Int. Conf. Comp.-Aided Des.* (2018), pp. 7:1–7:8

[Jia+18] S. Jiang et al., An efficient technique to reverse engineer minterm protection based camouflaged circuit. J. Comput. Sci. Technol. **33**(5), 998–1006 (2018)

[JM08] Y. Jin, Y. Makris, Hardware trojan detection using path delay fingerprint, in *Proc. Int. Symp. Hardw.-Orient. Sec. Trust* (2008), pp. 51–57. https://doi.org/10.1109/HST.2008.4559049

[Kar+18a] B. Karp et al., Detection and correction of malicious and natural faults in cryptographic modules, in *Proc. PROOFS@CHES* (2018), pp. 68–82. https://easychair.org/publications/download/zMjh

[Kar+18b] B. Karp et al., Security-oriented code-based architectures for mitigating fault attacks, in *Proc. DCIS* (2018), pp. 1–6. https://doi.org/10.1109/DCIS.2018.8681476

[Kes+18] S. Keshavarz et al., SAT-based reverse engineering of gate-level schematics using fault injection and probing, in *Proc. Int. Symp. Hardw.-Orient. Sec. Trust* (2018), pp. 215–220. https://doi.org/10.1109/HST.2018.8383918

[Kne+17] J. Knechtel et al., Large-scale 3D chips: challenges and solutions for design automation, testing, and trustworthy integration. Trans. Sys. LSI Des. Method. **10**, 45–62 (2017). https://doi.org/10.2197/ipsjtsldm.10.45

[Kne+19] J. Knechtel et al., Toward physically unclonable functions from plasmonics-enhanced silicon disc resonators. J. Lightwave Technol. **37**(15), 3805–3814 (2019). https://doi.org/10.1109/JLT.2019.2920949

[Koc+19a] P. Kocher et al., Spectre attacks: exploiting speculative execution, in *Proc. Symp. Sec. Priv.* , vol. 1, 19–37 (2019). https://doi.org/10.1109/SP.2019.00002. eprint: 1801.01203

[Kra+21] T. Krachenfels et al., Automatic extraction of secrets from the transistor jungle using laser-assisted side-channel attacks, in *Proc. USENIX Sec. Symp.* (2021)

[KV11] L.W. Kim, J.D. Villasenor, A system-on-chip bus architecture for thwarting integrated circuit trojan horses. Trans. VLSI Syst. **19**(10), 1921–1926 (2011). ISSN: 1063-8210. https://doi.org/10.1109/TVLSI.2010.2060375

[Lee+19] Y. Lee et al., Robust secure shield architecture for detection and protection against invasive attacks. Trans. Comput.-Aided Des. Integr. Circ. Syst. (2019). ISSN: 1937-4151. https://doi.org/10.1109/TCAD.2019.2944580

[Leo+19] J. Leonhard et al., MixLock: securing mixed-signal circuits via logic locking, in *Proc. Des. Autom. Test Europe* (2019)

[Ler+18] L. Lerman et al., Start simple and then refine: bias-variance decomposition as a diagnosis tool for leakage profiling. Trans. Comput. **67**(2), 268–283 (2018). https://doi.org/10.1109/TC.2017.2731342

[Li+16] M. Li et al., Provably secure camouflaging strategy for ic protection, in *Proc. Int. Conf. Comp.-Aided Des.* Austin, Texas (2016), pp. 28:1–28:8. ISBN: 978-1-4503-4466-1. https://doi.org/10.1145/2966986.2967065

[Li+17] M. Li et al., Provably secure camouflaging strategy for IC protection. Trans. Comput.-Aided Des. Integr. Circ. Syst. **38**, 1399 (2017). ISSN: 0278-0070. https://doi.org/10.1109/TCAD2017.2750088

[Li+18a] M. Li et al., A practical split manufacturing framework for trojan prevention via simultaneous wire lifting and cell insertion, in *Proc. Asia South Pac. Des. Autom. Conf.* (2018), pp. 265–270

[Li+18b] M. Li et al., TimingSAT: Decamouflaging timing-based logic obfuscation, in *Proc. Int. Test Conf.* (2018), pp. 1–10

[Li+19] H. Li et al., Attacking split manufacturing from a deep learning perspective, in *Proc. Des. Autom. Conf.* (2019), pp. 135:1–135:6. https://doi.org/10.1145/3316781.3317780

[Li+20] H. Li et al., Deep learning analysis for split manufactured layouts with routing perturbation. Trans. Comput.-Aided Des. Integr. Circ. Syst. **40**(10), 1995–2008 (2020).

[Lip+18a] M. Lipp et al., Meltdown. Comp. Research Rep. (2018). arXiv: 1801.01207. https://arxiv.org/abs/1801.01207

[Liu+14] C. Liu et al., Shielding heterogeneous MPSoCs from untrustworthy 3PIPs through security-driven task scheduling. Trans. Emerg. Top. Comput. **2**(4), 461–472 (2014). ISSN: 2168-6750. https://doi.org/10.1109/TETC.2014.2348182

[Liu+18] Y. Liu et al., A combined optimization-theoretic and side-channel approach for attacking strong physical unclonable functions. Trans. VLSI Syst. **26**(1), 73–81 (2018). ISSN: 1063-8210. https://doi.org/10.1109/TVLSI.2017.2759731

[LJM12] E. Love, Y. Jin, Y. Makris, Proof-carrying hardware intellectual property: a pathway to trusted module acquisition. Trans. Inf. Forens. Sec. **7**(1), 25–40 (2012). https://doi.org/10.1109/TIFS.2011.2160627

[LM06] H. Li, S. Moore, Security evaluation at design time against optical fault injection attacks. IEE Proc.-Inf. Sec. **153**(1), 3–11 (2006). https://doi.org/10.1049/ip-ifs:20055021

[LO19] L. Li, A. Orailoglu, Piercing logic locking keys through redundancy identification, in *Proc. Des. Autom. Test Europe* (2019). https://doi.org/10.23919/DATE.2019.8714955

[Loh+16] H. Lohrke et al., No place to hide: contactless probing of secret data on FPGAs, in *Proc. Cryptogr. Hardw. Embed. Sys.* (2016), pp. 147–167. ISBN: 978-3-662-53140-2. https://doi.org/10.1007/978-3-662-53140-2_8

[LP15] Y. Lao, K.K. Parhi, Obfuscating DSP circuits via high-level transformations. Trans. VLSI Syst. **23**(5), 819–830 (2015). ISSN: 1063-8210. 1.1109/TVLSI.2014.2323976

[Mae+18] P. Maene et al., Hardware-based trusted computing architectures for isolation and attestation. Trans. Comput. **67**(3), 361–374 (2018). ISSN: 0018-9340. https://doi.org/10.1109/TC.2017.2647955

[Man+19] S. Manipatruni et al., Scalable energy-efficient magnetoelectric spin–orbit logic. Nature **565**(7737), 35–42 (2019)

[McC16] C. McCants. *Trusted Integrated Chips (TIC) Program* (2016). https://www.ndia.org/-/media/sites/ndia/meetings-and-events/divisions/systems-engineering/past-events/trusted-micro/2016-august/mccants-carl.ashx

[MGT15] M. El Massad, S. Garg, M.V. Tripunitara, Integrated circuit (IC) decamouflaging: reverse engineering camouflaged ICs within minutes, in *Proc. Netw. Dist. Sys. Sec. Symp.* (2015), pp. 1–14. https://doi.org/10.14722/ndss.2015.23218

[MSD16] J. Magaña, D. Shi, A. Davoodi, Are Proximity attacks a threat to the security of split manufacturing of integrated circuits?, in *Proc. Int. Conf. Comp.-Aided Des.* Austin, Texas (2016), pp. 90:1–90:7. ISBN: 978-1-4503-4466-1. https://doi.org/10.1145/2966986.2967006

[MV10] R. Maes, I. Verbauwhede, Physically unclonable functions: a study on the state of the art and future research directions, in *Towards Hardware-Intrinsic Security: Foundations and Practice*, ed. by A.-R. Sadeghi, D. Naccache (Springer, Berlin, 2010), pp. 3–37. ISBN: 978-3-642-14452-3. https://doi.org/10.1007/978-3-642-14452-3_1

[Nab+20] M. Nabeel et al., 2.5D root of trust: secure system-level integration of untrusted chiplets. Trans. Comput. **69**(11), 1611–1625 (2020). https://doi.org/10.1109/TC.2020.3020777. Dedicated, after acceptance and publication, in memory of the late Vassos Soteriou; version with dedication note available at https://arxiv.org/abs/2009.02412

[Nat+19] G. Di Natale et al., Hidden-delay-fault sensor for test, reliability and security, in *Proc. Des. Autom. Test Europe* (2019), pp. 316–319. https://doi.org/10.23919/DATE.2019.8714891

[Nir+16] I.R. Nirmala et al., A novel threshold voltage defined switch for circuit camouflaging, in *Proc. Europe Test. Symp.* (2016), pp. 1–2. https://doi.org/10.1109/ETS.2016.7519286

[OD19] C. O'Flynn, A. Dewar, On-device power analysis across hardware security domains. *Trans. Cryptogr. Hardw. Embed. Syst.* **2019**(4), 126–153 (2019). https://doi.org/10.13154/tches.v2019.i4.126-153

[OST05] D.A. Osvik, A. Shamir, E. Tromer, Cache attacks and countermeasures: the case of AES, in *IACR Crypt. ePrint Arch.*, vol. 271 (2005). https://eprint.iacr.org/2005/271

[Pan+19] J. Pan et al., One fault is all it needs: breaking higher-order masking with persistent fault analysis, in *Proc. Des. Autom. Test Europe* (2019), pp. 1–6. ISBN 978-3-9819263-2-3

[Pap+02] R. Pappu et al., Physical one-way functions. Science **297**(5589), 2026–2030 (2002). ISSN: 0036-8075. https://doi.org/10.1126/science.1074376. eprint: http://science.sciencemag.org/content/297/5589/2026.full.pdf

[Par+17] F. Parveen et al., Hybrid polymorphic logic gate with 5-terminal magnetic domain wall motion device, in *Proc. Comp. Soc. Symp. VLSI* (2017), pp. 152–157. https://doi.org/10.1109/ISVLSI.2017.35

[Pat+17] S. Patnaik et al., Obfuscating the Interconnects: low-cost and resilient full-chip layout camouflaging, in *Proc. Int. Conf. Comp.-Aided Des.* (2017), pp. 41–48. https://doi.org/10.1109/ICCAD.2017.8203758

[Pat+18c] S. Patnaik et al., Best of both worlds: integration of split manufacturing and camouflaging into a security-driven CAD flow for 3D ICs, in *Proc. Int. Conf. Comp.-Aided Des.* (2018)

[Pat+18e] S. Patnaik et al., Concerted wire lifting: enabling secure and cost-effective split manufacturing, in *Proc. Asia South Pac. Des. Autom. Conf.* (2018), pp. 251–258. https://doi.org/10.1109/ASPDAC.2018.8297314

[Pat+18f] S. Patnaik et al., Raise your game for split manufacturing: restoring the true functionality through BEOL, in *Proc. Des. Autom. Conf.* (2018), pp. 140:1–140:6. https://doi.org/10.1145/3195970.3196100

[Pat+19a] S. Patnaik et al., A modern approach to IP protection and trojan prevention: split manufacturing for 3D ICs and obfuscation of vertical interconnects, in *IEEE Transactions on Emerging Topics in Computing* (2019)

[Pat+19b] S. Patnaik et al., A modern approach to IP protection and trojan prevention: split manufacturing for 3D ICs and obfuscation of vertical interconnects. Trans. Emerg. Top. Comput. Early Access (2019). https://doi.org/10.1109/TETC.2019.2933572

[Pat+20a] S. Patnaik et al., Obfuscating the interconnects: low-cost and resilient full-chip layout camouflaging. Trans. Comput. Aided Des. Integr. Circ. Syst. **39**(12), 4466–4481 (2020). https://doi.org/10.1109/TCAD.2020.2981034

[Pat+21] S. Patnaik et al., Concerted wire lifting: Enabling secure and cost-effective split manufacturing. Trans. Comput. Aided Des. Integr. Circ. Syst. (2021) https://doi.org/10.1109/TCAD.2021.3056379

[PM15] S.M. Plaza, I.L. Markov, Solving the third-shift problem in IC piracy with test-aware logic locking. Trans. Comput. Aided Des. Integr. Circ. Syst. **34**(6), 961–971 (2015)

[Pri+17a] E.L. Principe et al., Plasma FIB deprocessing of integrated circuits from the backside. Elec. Dev. Fail. Anal. **19**(4), 36–44 (2017). https://www.researchgate.net/profile/Robert_Chivas/publication/322264562_Plasma_FIB_deprocessing_of_integrated_circuits_from_the_backside/links/5a54f88e45851547b1bd55f2/Plasma-FIB-deprocessing-of-integrated-circuits-from-the-backside.pdf

[Qiu+19] P. Qiu et al., VoltJockey: Breaching TrustZone by software-controlled voltage manipulation over multi-core frequencies, in *Proc. Comp. Comm. Sec.* (2019), pp. 195–209. ISBN: 978-1-4503-6747-9. https://doi.org/10.1145/3319535.3354201

[Rai+18] S. Rai et al., Emerging reconfigurable nanotechnologies: Can they support future electronics?, in *2018 IEEE/ACM International Conference on Computer-Aided Design (IC-CAD)* (IEEE, Piscataway, 2018), pp. 1–8

[Raj+12] J. Rajendran et al., Security analysis of logic obfuscation, in *Proc. Des. Autom. Conf.* (2012)

[Raj+13a] J. Rajendran et al., Security analysis of integrated circuit camouflaging, in *Proc. Comp. Comm. Sec.* Berlin, Germany (2013), pp. 709–720. ISBN: 978-1-4503-2477-9. https://doi.org/10.1145/2508859.2516656

[Raj+15] J. Rajendran et al., Fault analysis-based logic encryption. Trans. Comput. **64**(2), 410–424 (2015)

[Ram19] Rambus Inc. *Circuit Camouflage Technology* (2019). https://www.rambus.com/security/cryptofirewall-cores/circuit-camouflage-technology/

[Ran+19] N. Rangarajan et al., Spin-based reconfigurable logic for power-and area-efficient applications. IEEE Des. Test **36**(3), 22–30 (2019)

[Ran+20a] N. Rangarajan et al., Opening the doors to dynamic camouflaging: harnessing the power of polymorphic devices. Trans. Emerg. Top. Comput. Early Access (2020). https://doi.org/10.1109/TETC.2020.2991134

[RKM10] J.A. Roy, F. Koushanfar, I.L. Markov, Ending piracy of integrated circuits. Computer **43**(10), 30–38 (2010). ISSN: 0018-9162. https://doi.org/10.1109/MC.2010.284

[RMS18] S. Roshanisefat, H. Mardani Kamali, A. Sasan, SRCLock: SAT-resistant cyclic logic locking for protecting the hardware, in *Proc. Great Lakes Symp. VLSI* (2018), pp. 153–158

[RSK13] J. Rajendran, O. Sinanoglu, R. Karri, Is split manufacturing secure?, in *Proc. Des. Autom. Test Europe* (2013), pp. 1259–1264. https://doi.org/10.7873/DATE.2013.261

[Rüh+13a] U. Rührmair et al., PUF modeling attacks on simulated and silicon data. Trans. Inf. Forens. Sec. **8**(11), 1876–1891 (2013). ISSN: 1556-6013. https://doi.org/10.1109/TIFS.2013.2279798

[Rüh+13b] U. Rührmair et al., Optical PUFs reloaded, in *IACR Crypt. ePrint Arch.* (2013). https://eprint.iacr.org/2013/215

[Sch+19] M. Schwarz et al., ZombieLoad: cross-privilege-boundary data sampling. Comp. Research Rep. (2019). https://arxiv.org/abs/1905.05726

[Sen+17] A. Sengupta et al., Rethinking split manufacturing: an information-theoretic approach with secure layout techniquesy, in *Proc. Int. Conf. Comp.-Aided Des.* (2017), pp. 329–336. https://doi.org/10.1109/ICCAD.2017.8203796. Revised version available at https://arxiv.org/abs/1710.02026

[Sen+19] A. Sengupta et al., A new paradigm in split manufacturing: lock the FEOL, unlock at the BEOL, in *Proc. Des. Autom. Test Europe* (2019)

[Sen+20] A. Sengupta et al., Truly stripping functionality for logic locking: A fault-based perspective. Trans. Comput. Aided Des. Integr. Circ. Syst. **39**(12), 4439–4452 (2020)

[Sha+17a] K. Shamsi et al., AppSAT: approximately deobfuscating integrated circuits, in *Proc. Int. Symp. Hardw.-Orient. Sec. Trust* (2017), pp. 95–100. https://doi.org/10.1109/HST. 2017.7951805

[Sha+17b] K. Shamsi et al., Cyclic obfuscation for creating SAT-unresolvable circuits, in *Proc. Great Lakes Symp. VLSI*. Banff, Alberta (2017), pp. 173–178. ISBN: 978-1-4503-4972-7. https://doi.org/10.1145/3060403.3060458

[Sha+18a] K. Shamsi et al., On the approximation resiliency of logic locking and IC camouflaging schemes. Trans. Inf. Forens. Sec. (2018). ISSN: 1556-6013. https://doi.org/10. 1109/TIFS.2018.2850319

[Sha+18b] K. Shamsi et al., Cross-lock: dense layout-level interconnect locking using cross-bar architectures, in *Proc. Great Lakes Symp. VLSI* (2018), pp. 147–152

[Sha+20] M.M. Sharifi et al., A novel TIGFET-based DFF design for improved resilience to power side-channel attacks, in *Proceedings of the 23rd Conference on Design, Automation and Test in Europe* (2020), pp. 1253–1258

[She+18] H. Shen et al., Nanopyramid: an optical scrambler against backside probing attacks, in *Proc. Int. Symp. Test. Failure Analys.* (2018). https://pdfs.semanticscholar.org/453a/ ce0749c374d59c4193cc26d06ac38e22c500.pdf

[She+19] Y. Shen et al., BeSAT: behavioral SAT-based attack on cyclic logic encryption, in *Proc. Asia South Pac. Des. Autom. Conf.* (2019), pp. 657–662

[Shi+17] Q. Shi et al., Securing split manufactured ICs with wire lifting obfuscated built-in self-authentication, in *Proc. Great Lakes Symp. VLSI*. Banff, Alberta (2017), pp. 339–344. ISBN: 978-1-4503-4972-7. https://doi.org/10.1145/3060403.3060588

[SHS16] B. Selmke, J. Heyszl, G. Sigl, Attack on a DFA protected AES by simultaneous laser fault injections, in *Proc. Worksh. Fault Diag. Tol. Cryptogr.* (2016), pp. 36–46. https:// doi.org/10.10.1109/FDTC.2016.16

[SR07] M. Stipčević, B.M. Rogina, Quantum random number generator based on photonic emission in semiconductors. Rev. Sci. Instrum. **78**(4), 045104 (2007)

[SRM15] P. Subramanyan, S. Ray, S. Malik, Evaluating the security of logic encryption algorithms, in *Proc. Int. Symp. Hardw.-Orient. Sec. Trust* (2015), pp. 137–143. https:// doi.org/10.1109/HST.2015.7140252

[SS19] D. Sirone, P. Subramanyan, Functional analysis attacks on logic locking, in *Proc. Des. Autom. Test Europe* (2019)

[Sug+15] T. Sugawara et al., Reversing stealthy dopant-level circuits. J. Cryptogr. Eng. **5**(2), 85–94 (2015). ISSN: 2190-8516. https://doi.org/10.1007/s13389-015-0102-5

[SW12] S. Skorobogatov, C. Woods, In the blink of an eye: There goes your AES key, in *IACR Crypt. ePrint Arch.*, vol. 296 (2012)

[Swi+17] P. Swierczynski et al., Interdiction in practice–hardware trojan against a high-security USB flash drive. J. Cryptogr. Eng. **7**(3), 199–211 (2017). ISSN: 2190-8516. https:// doi.org/10.1007/s13389-016-0132-7

[SZ17] Y. Shen, H. Zhou, Double DIP: re-evaluating security of logic encryption algorithms, in *Proc. Great Lakes Symp. VLSI* (2017), pp. 179–184. https://doi.org/10.1145/ 3060403.3060469

[Taj+17] S. Tajik et al., On the power of optical contactless probing: attacking bitstream encryption of FPGAs, in *Proc. Comp. Comm. Sec.* (2017), pp. 1661–1674. https:// doi.org/10.1145/3133956.3134039

[TŠ07] P. Tuyls, B. Škorić, Strong authentication with physical unclonable functions, in *Security, Privacy, and Trust in Modern Data Management*, ed. by M. Petković, W. Jonker (Springer, Berlin, 2007), pp. 133–148. ISBN: 978-3-540-69861-6. https://doi. org/10.1007/978-3-540-69861-6_10

[Vai+14a] K. Vaidyanathan et al., Building trusted ICs using split fabrication, in *Proc. Int. Symp. Hardw.-Orient. Sec. Trust* (2014), pp. 1–6. https://doi.org/10.1109/HST.2014. 6855559

[Vai+14b] K. Vaidyanathan et al., Efficient and secure intellectual property (IP) design with split fabrication, in *Proc. Int. Symp. Hardw.-Orient. Sec. Trust* (2014), pp. 13–18. https:// doi.org/10.1109/HST.2014.6855561

[Val+13] J. Valamehr et al., A 3-D split manufacturing approach to trustworthy system development. Trans. Comput. Aided Des. Integr. Circ. Syst. **32**(4), 611–615 (2013). ISSN: 0278-0070. https://doi.org/10.1109/TCAD.2012.2227257

[Vas+18] N. Vashistha et al., Trojan scanner: detecting hardware trojans with rapid SEM imaging combined with image processing and machine learning, in *Proc. Int. Symp. Test. Failure Analys.* (2018). https://pdfs.semanticscholar.org/7b7d/ 582034c19096c28c47bd1452e8becf287abc.pdf

[VDP14] K. Vaidyanathan, B.P. Das, L. Pileggi, Detecting Reliability attacks during split fabrication using test-only BEOL stack, in *Proc. Des. Autom. Conf.* San Francisco, CA (2014), pp. 156:1–156:6. ISBN: 978-1-4503-2730-5. https://doi.org/10.1145/ 2593069.2593123

[Vee+16] V. van der Veen et al., Drammer: deterministic Rowhammer attacks on mobile platforms, in *Proc. Comp. Comm. Sec.* Vienna (2016), pp. 1675–1689. ISBN: 978-1-4503-4139-4. https://doi.org/10.1145/2976749.2978406

[Ven+15] R. Venkatesan et al., Spintastic: Spin-based stochastic logic for energy-efficient computing, in *2015 Design, Automation & Test in Europe Conference & Exhibition (DATE)* (IEEE, Piscataway, 2015), pp. 1575–1578

[Wah+16] R.S. Wahby et al., Verifiable ASICs, in *Proc. Symp. Sec. Priv.* (2016), pp. 759–778. https://doi.org/10.1109/SP.2016.51

[Wan+15b] X. Wang et al., IIPS: Infrastructure IP for secure SoC design. Trans. Comput. **64**(8), 2226–2238 (2015). ISSN: 0018-9340. https://doi.org/10.1109/TC.2014.2360535

[Wan+16a] X. Wang et al., Secure and low-overhead circuit obfuscation technique with multiplexers, in *Proc. Great Lakes Symp. VLSI*. Boston, Massachusetts (2016), pp. 133–136. ISBN: 978-1-4503-4274-2. https://doi.org/10.1145/2902961.2903000

[Wan+16b] Y. Wang et al., The cat and mouse in split manufacturing, in *Proc. Des. Autom. Conf.* Austin, Texas (2016), pp. 165:1–165:6. ISBN: 978-1-4503-4236-0. https://doi.org/10. 1145/2897937.2898104

[Wan+17a] H. Wang et al., Probing attacks on integrated circuits: challenges and research opportunities. Des. Test **34**(5), 63–71 (2017). ISSN: 2168-2356. https://doi.org/10. 1109/MDAT.2017.2729398

[Wan+17c] Y. Wang et al., Front-end-of-line attacks in split manufacturing, in *Proc. Int. Conf. Comp.-Aided Des.* (2017)

[Wan+17d] Y. Wang et al., Routing perturbation for enhanced security in split manufacturing, in *Proc. Asia South Pac. Des. Autom. Conf.* (2017), pp. 605–610. https://doi.org/10. 1109/ASPDAC.2017.7858390

[Wan+18a] X. Wang et al., Parallelizing SAT-based de-camouflaging attacks by circuit partitioning and conflict avoiding. Integration **67**, 108 (2018). ISSN: 0167-9260. https://doi. org/10.1016/j.vlsi.2018.10.009

[Wan+19] H. Wang et al., Probing assessment framework and evaluation of antiprobing solutions. Trans. VLSI Syst. **27**(6), 1239–1252 (2019). ISSN: 1557-9999. https://doi.org/ 10.1109/TVLSI.2019.2901449

[Wei+18] M. Weiner et al., The low area probing detector as a countermeasure against invasive attacks. Trans. VLSI Syst. **26**(2), 392–403 (2018). ISSN: 1063-8210. https://doi.org/ 10.1109/TVLSI.2017.2762630

[Wu+16] T.F. Wu et al., TPAD: hardware trojan prevention and detection for trusted integrated circuits. Trans. Comput. Aided Des. Integr. Circ. Syst. **35**(4), 521–534 (2016). ISSN: 0278-0070. https://doi.org/10.1109/TCAD.2015.2474373

[XFT14] K. Xiao, D. Forte, M. Tehranipoor, A novel built-in self-authentication technique to prevent inserting hardware trojans. Trans. Comput. Aided Des. Integr. Circ. Syst. **33**(12), 1778–1791 (2014). ISSN: 0278-0070. https://doi.org/10.1109/TCAD.2014.2356453

[XFT15] K. Xiao, D. Forte, M.M. Tehranipoor, Efficient and secure split manufacturing via obfuscated built-in self-authentication, in *Proc. Int. Symp. Hardw.-Orient. Sec. Trust* (2015), pp. 14–19. https://doi.org/10.1109/HST.2015.7140229

[Xia+09] Q. Xia et al., Memristor- CMOS hybrid integrated circuits for reconfigurable logic. Nano Lett. **9**(10), 3640–3645 (2009)

[XS16] Y. Xie, A. Srivastava, Mitigating SAT attack on logic locking, in *Proc. Cryptogr. Hardw. Embed. Sys.* (2016), pp. 127–146. https://doi.org/10.1007/978-3-662-53140-2_7

[XS17] Y. Xie, A. Srivastava, Delay locking: Security enhancement of logic locking against IC counterfeiting and over-production, in *Proc. Des. Autom. Conf.* (2017), pp. 9:1–9:6

[Xu+17] X. Xu et al., Novel bypass attack and BDD-based trade-off analysis against all known logic locking attacks, in *Proc. Cryptogr. Hardw. Embed. Sys.* (2017)

[Xu+19] W. Xu et al., Layout recognition attacks on split manufacturing, in *Proc. Asia South Pac. Des. Autom. Conf.* (2019), pp. 45–50. https://doi.org/10.1145/3287624.3287698

[Yas+16a] M. Yasin et al., Activation of logic encrypted chips: Pre-test or post-test?, in *Proc. Des. Autom. Test Europe* (2016), pp. 139–144

[Yas+16b] M. Yasin et al., SARLock: SAT attack resistant logic locking, in *Proc. Int. Symp. Hardw.-Orient. Sec. Trust* (2016). https://doi.org/10.1109/HST.2016.7495588

[Yas+16c] M. Yasin et al., Security analysis of anti-SAT, in *Proc. Asia South Pac. Des. Autom. Conf.* (2016), pp. 342–347

[Yas+16d] M. Yasin et al., CamoPerturb: secure IC camouflaging for minterm protection, in *Proc. Int. Conf. Comp.-Aided Des.* (2016), pp. 29:1–29:8. https://doi.org/10.1145/2966986.2967012

[Yas+17a] M. Yasin et al., What to lock?: Functional and parametric locking, in *Proc. Great Lakes Symp. VLSI* (2017), pp. 351–356

[Yas+17b] M. Yasin et al., Provably-secure logic locking: from theory to practice, in *Proc. Comp. Comm. Sec.* (2017), pp. 1601–1618. https://doi.org/10.1145/3133956.3133985

[Yas+17c] M. Yasin et al., Removal attacks on logic locking and camouflaging techniques. Trans. Emerg. Top. Comput. **8**, 517 (2017)

[YPK16] K. Yi, M. Park, S. Kim, Practical silicon-surface-protection method using metal layer. J. Semicond. Tech. Sci. **16**(4), 470–480 (2016). https://doi.org/10.5573/JSTS.2016.16.4.470

[YRS20] M. Yasin, J. (JV) Rajendran, O. Sinanoglu, *Trustworthy Hardware Design: Combinational Logic Locking Techniques* (Springer, Cham, 2020). https://doi.org/10.1007/978-3-030-15334-2

[YS17] M. Yasin, O. Sinanoglu, Evolution of logic locking, in *Proc. VLSI SoC* (2017). https://doi.org/10.1109/VLSI-SoC-2017.8203496

[YSR17] M. Yasin, O. Sinanoglu, J. Rajendran, Testing the trust-worthiness of IC testing: an oracle-less attack on IC camouflaging. Trans. Inf. Forens. Sec. **12**(11), 2668–2682 (2017). ISSN: 1556-6013. https://doi.org/10.1109/TIFS.2017.2710954

[YTS19] F. Yang, M. Tang, O. Sinanoglu, Stripped functionality logic locking with hamming distance based restore unit (SFLL-hd) – unlocked. Trans. Inf. Forens. Sec. (2019). https://doi.org/10.1109/TIFS.2019.2904838

[Yu+17] C. Yu et al., Incremental SAT-based reverse engineering of camouflaged logic circuits. Trans. Comput. Aided Des. Integr. Circ. Syst. **36**(10), 1647–1659 (2017). ISSN: 0278-0070. https://doi.org/10.1109/TCAD.2017.2652220

[Zam+18] M. Zaman et al., Towards provably-secure performance locking, in *Proc. Des. Autom. Test Europe* (2018), pp. 1592–1597

[ZF05] Y.B. Zhou, D.G. Feng, Side-channel attacks: ten years after its publication and the impacts on cryptographic module security testing, in *IACR Crypt. ePrint Arch.*, vol. 388 (2005). http://eprint.iacr.org/2005/388

[Zha+18a] G.L. Zhang et al., TimingCamouflage: Improving circuit security against counterfeiting by unconventional timing, in *Proc. Des. Autom. Test Europe* (2018), pp. 91–96. https://doi.org/10.23919/DATE.2018.8341985

[Zha+18b] R. Zhang et al., Nanoscale diffusive memristor crossbars as physical unclonable functions. Nanoscale **10**(6), 2721–2726 (2018)

[Zha+19] H. Zhang et al., Architectural support for containment-based security, in *Proc. Arch. Supp. Programm. Lang. Op. Sys.* (2019), pp. 361–377. https://doi.org/10.1145/3297858.3304020

[ZJK17] H. Zhou, R. Jiang, S. Kong, CycSAT: SAT-based attack on cyclic logic encryptions, in *Proc. Int. Conf. Comp.-Aided Des.* (2017), pp. 49–56

[ZMD18] B. Zhang, J.C. Magaña, A. Davoodi, Analysis of security of split manufacturing using machine learning, in *Proc. Des. Autom. Conf.* (2018), pp. 141:1–141:6

Chapter 2
Reconfigurability for Static Camouflaging

2.1 Chapter Introduction

Intellectual property (IP) piracy in the modern integrated circuit (IC) supply chain has necessitated countermeasures like layout camouflaging to safeguard proprietary design IP. Although effective initially, the conventional CMOS-centric layout camouflaging primitives hardly challenge attackers having access to advanced reverse engineering tools. Equipped with these state-of-the-art tools, which include de-layering and de-packaging equipment, scanning electron microscopy, and Boolean satisfiability (SAT)-based attacks, attackers can unmask the functionality of a camouflaged chip in minutes [EGT15]. The substantial overheads associated with CMOS-based camouflaging essentially limit its pervasive use across the chip [Raj+13b]. Emerging technologies could be a promising solution to this problem and enable the protection of a considerably larger portion of the design IP.

Static camouflaging, in particular, is achieved by reconfigurable emerging technologies that exhibit multiple device functionalities in the pre-fabrication stages. As the name implies, static camouflaging requires a device that can realize several Boolean functions but is programmed to implement only one of them at a time. Once set, the functionality remains static but unknown to the attacker since all the possible functions are accomplished with the same device layout. This contrasts with dynamic camouflaging, which requires a device capable of morphing its functionality multiple times on-the-fly. Static camouflaging is the preferred protection technique for high-performance deterministic processing systems, whereas dynamic camouflaging caters more to approximate computational circuits.

In this chapter, we present the fundamental concepts and considerations for constructing a static camouflaging primitive with emerging devices. We then examine three such devices and study how they were used to devise such a primitive. We also demonstrate a case study of an emerging device-based static camouflaging scheme, detailing the steps involved in the process, from conception to circuit implementation, and finally to security analysis.

© The Author(s), under exclusive license to Springer Nature Switzerland AG 2021
N. Rangarajan et al., *The Next Era in Hardware Security*,
https://doi.org/10.1007/978-3-030-85792-9_2

2.2 Concepts for Static Camouflaging Using Reconfigurable Devices

Designing a secure and robust static camouflaging primitive hinges on selecting an emerging device, which can achieve reconfigurability. Reconfigurability, in this context, refers to the ability of a particular device to be configured as multiple logic functions *without* changing the underlying structure of the device. In the case of static camouflaging, this reconfigurability is only required in the pre-fabrication stages of the supply chain. We explore post-fabrication reconfigurability on-the-fly for dynamic camouflaging in the subsequent chapter. The important considerations for choosing a reconfigurable emerging device for static camouflaging are as follows:

1. **Multi-functional** devices can realize several logic functions using one topology, wherein a control signal is typically used to program the functionality. Since the internal details of this control signal are unknown to the attacker, the functionality remains hidden. In the case of static camouflaging, the control signal is programmed once during the fabrication or testing phase and remains static throughout the lifetime of the circuit. However, the control signal could be programmed to one of the multiple possible states, resulting in different logic gates.
2. For instance, several emerging logic technologies possess the capability to directly implement **Majority (MAJ) logic** in a single device, a feat that requires multiple gates in CMOS logic. By utilizing one of its inputs as a tie-breaking control input, an odd-input (emerging device-based) MAJ gate can be tuned to function as AND or OR. If the emerging device has intrinsic inversion, controllable via an external terminal, then NAND and NOR can be achieved as well.
3. Some emerging device technologies enable the construction of **optically indistinguishable** circuit elements and gates, wherein an attacker cannot differentiate between the elements via optical imaging or inspection. Such devices generally produce identical layouts for the different gates implemented.
4. Inserting **dummy contacts** is a secure design practice, which allows the designer to make multi-functional devices/gates optically indistinguishable at the layout level. The dummy contacts cover up any physical discrepancies in the wiring connections for the different versions of the multi-functional device, thus making them uniform. However, advanced imaging techniques that can detect the charge flow through contacts can decipher dummy contacts from real ones.
5. Optically indistinguishable circuit elements are often forged from **ambipolar** emerging devices that can change their threshold voltage and majority carriers, by manipulating the channel doping. Since the doping level is not easily decoded by an attacker, the functionality remains hidden.

2.3   Review of Selected Emerging Technologies and Prior Art

In this section, we briefly review prior static camouflaging schemes, constructed with various classes of emerging devices and phenomena. This survey is intended to provide the reader with insights into how emerging multi-functional devices can be exploited to tackle IP piracy.

2.3.1   All-Spin Logic

The all-spin logic (ASL), introduced by Behtash et al. [Beh+10], was leveraged to build a static camouflaging primitive in [AYF17]. The ASL device utilizes non-local spin signals from nanomagnetic switching, to drive logic. The proposed camouflaging scheme can implement multi-functional ASL structures capable of realizing either NAND/NOR/AND/OR or XOR/XNOR or INV/BUF gates.

The basic ASL gate, shown in Fig. 2.1, consists of (1) two bistable magnetic free layers (input and output magnets), whose magnetization defines the binary state, (2) an isolation layer to separate different stages of the logic, (3) a non-magnetic channel to transport the spin signals from the input magnet to the output magnet, (4) a tunneling layer for the spins to tunnel from the input magnet into the channel, and (5) contacts to connect the supply and the ground. Here, the input and output ports of the magnets have low and high spin polarization factors, respectively. The channel is composed of a high spin-coherence length material like Cu.

The working of the basic ASL gate is as follows. A negative voltage on the input terminal will result in the injection of majority spins, parallel to the input magnet's state, into the channel. This spin current switches the output magnet parallel to the input magnet, thus implementing the buffer (BUF) operation, whereas a positive

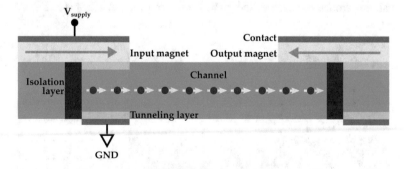

Fig. 2.1 Construction of the basic ASL INV/BUF gate. A negative supply voltage results in the BUF operation, whereas a positive supply voltage implements an INV

voltage on the input terminal extracts majority spins from the channel, resulting in an opposite spin current that switches the output magnet anti-parallel with respect to the input magnet. This corresponds to the inversion (INV) operation. For further details on the construction and working of the ASL device, readers are referred to [Beh+10].

Majority logic is implemented in ASL by connecting several input terminals to one output terminal, wherein the output is dominated by the majority of the inputs. The authors in [AYF17] use a 3-input MAJ ASL structure with two primary inputs and one programmed tie-breaking key to realize complex Boolean gates. Here a key input of 0 implements the AND (positive supply voltage) and NAND (negative supply voltage) functions, whereas a key input of 1 results in OR (positive supply voltage) and NOR (negative supply voltage) gates. XOR and XNOR gates can be achieved in a 5-Terminal ASL structure through a sum of Full Adders operation, details of which can be found in [Aug+11].

Alasad et al. [AYF17] utilize this library of multi-functional ASL gates to demonstrate static camouflaging against brute force key search. However, their resilience against advanced SAT attacks is questionable, due to the small number of functions implemented with identical device layouts. Further, the relatively high energy overheads due to the ASL device impede the large-scale use of this scheme on real-world circuits.

2.3.2 Silicon Nanowire Transistors

Silicon nanowire field effect transistors (SiNWFET) are viable devices for static camouflaging owing to their ambipolarity. As demonstrated by Bi et al. [Bi+14], the ability to switch the majority carrier type from n to p under normal bias conditions enables the reconfiguration of the device polarity to achieve multi-functionality. As illustrated in Fig. 2.2, they consider a vertically stacked SiNWFET structure with two gate-all around electrodes, the control gate and the polarity gate. The control

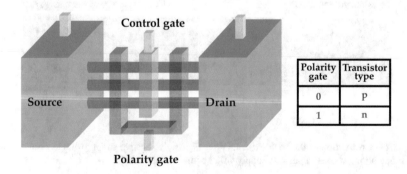

Polarity gate	Transistor type
0	p
1	n

Fig. 2.2 Ambipolar SiNWFET, where the polarity gate is used to tune the transistor type

gate is the conventional gate terminal, which switches the channel ON/OFF. The polarity gate, placed near the source/drain Schottky junctions, allows one to tune the device polarity.

Bi et al. propose a four SiNWFET layout for implementing NAND/NOR/XOR/ XNOR functionalities, with different real and dummy contact configurations. This layout, detailed in [Bi+14], greatly reduces the area and power overheads as compared to similar camouflaging schemes in CMOS, which would require at least 12 transistors. However, the small number of functions implemented with the universal layout again hinders SAT resilience. This, in turn, would require a larger number of gates to be camouflaged, thus increasing overhead costs.

2.3.3   Two-Dimensional Devices

Wali et al. [Wal+21] proposed an advanced static camouflaging scheme by harnessing two-dimensional (2D) transition metal dichalcogenides (TMD) and transition metal oxides (TMO). The basic premise of their work hinges on the unique property of TMDs, wherein they can be molded into various circuit components, including n- and p-type transistors, diodes, and resistors, using optically indistinguishable device structures. Plasma oxidation of the deposited TMD causes the oxidation of the top layers, resulting in substoichiometric TMOs that are optically transparent. This leaves behind a TMO/TMD heterostructure, which is partially transparent from the top, and resilient to optical inspection and analysis-based reverse engineering (RE). In fact, optical imaging-based techniques do not yield any information about the TMO thickness, plasma exposure time, or distinguishing color spectra in the TMO layers.

Their camouflaged resistor consists of a TMD/TMO heterostructure (Fig. 2.3a), which can implement various resistance values using the same device footprint and structure. The resistance can be adjusted by simply varying the plasma exposure time, which results in TMO layers of different thicknesses. However, the TMO being transparent is able to conceal this thickness and hence its resistance, from any attacker. The camouflaged diode is constructed by temporarily covering one side of the resistor heterostack by poly(methyl methacrylate), before introducing the oxygen plasma. This results in a diode junction, since the covered area remains intrinsic, while the exposed area is doped p-type owing to the substoichiometric TMO formation. Camouflaged transistors are achieved by leveraging the ambipolarity in the TMD/TMO heterostructure. As the plasma exposure time is increased, the heterostructure undergoes a change from n-type to p-type. This transformation does not leave any optical trace and both the n-type and p-type 2D-FETs are indistinguishable.

The camouflaged circuit elements are used in a black box configuration as shown in Fig. 2.3b, to realize various logic gates. Since the elements are identical, the gates constructed with them end up indistinguishable as well. The negligible overheads due to the intrinsic camouflaging property of these 2D heterostructures enable

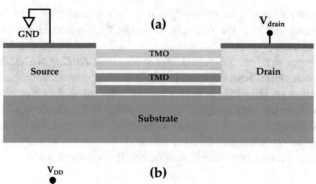

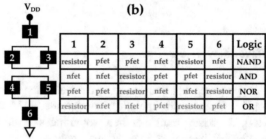

Fig. 2.3 (**a**) Schematic of the 2D TMD/TMO heterostructure device to implement a camouflaged resistor. The diode and FET have an additional body terminal. (**b**) Construction of camouflaged logic gates

100% camouflaging of all gates in the target circuit. This thwarts any brute force attack, and even advance SAT attacks are mitigated due to the increased attack time complexity. Further, automatic test pattern generation (ATPG) attacks, which rely on the targeted activation of specific gates and the subsequent propagation of their local outputs to the primary outputs, are rendered futile against this 2D device-based camouflaging scheme. Note however that advanced RE setups employing a combination of Raman spectroscopy, atomic force microscopy, and scanning electron microscopy may be able to discern differences in the TMO layers, which could simplify the search space for RE.

2.4   Case Study: Static Camouflaging Using Giant Spin Hall Effect Devices

In [Pat+18b], Patnaik et al. proposed for the first time, a static camouflaging primitive that is capable of implementing all 16 possible Boolean functionalities for two inputs. With such a large solution space for the attacker to cover, even a modest 30–40% of circuit camouflaging renders SAT-based attacks computationally unfeasible. In this case study, we explore the different aspects and details of this primitive, to grasp the various steps involved in the design, implementation, and testing of a static camouflaging scheme.

The giant spin Hall effect (GSHE) switch, which is at the heart of the proposed primitive, is constructed by combining a heavy-metal spin Hall layer, such as tantalum, tungsten, platinum, or palladium, with a magnetic tunnel junction (MTJ) arrangement (Fig. 2.4). Above the heavy-metal layer are two nanomagnets for write and read modes (W-NM and R-NM, red). The W-NM is separated from the output terminal via an insulating oxide layer (green). On top of the R-NM sit two fixed ferromagnetic layers (dark green) with anti-parallel magnetization directions.

The switch relies on the spin Hall effect [Hir99] for generating and amplifying the spin current input, and the magnetic dipolar coupling phenomenon [Kan+15] to magnetically couple R-NM and W-NM, while keeping them electrically isolated. A charge current through the bottom heavy-metal layer (along $\hat{x}$) induces a spin current in the transverse direction (along $\hat{y}$), which is used to switch the magnetization state of the W-NM. The dipolar coupling field then causes the R-NM to switch its orientation. That is because in the presence of magnetic dipolar coupling, the minimum energy state is the one in which the W-NM and the R-NM are anti-parallel to each other [DSB12]. The final magnetization state of the R-NM is read off using a differential MTJ setup. The logic (1 or 0) is encoded in the direction of the electrical output current (+I or −I). The current direction depends on the relative orientations of the fixed magnets in the MTJ stack with respect to the final magnetization of the R-NM. The parallel path offers a lower resistance for a charge current passing either from the MTJ contact to the output terminal or vice versa (i.e., from the output terminal to the MTJ contact). Hence, depending on the polarity of

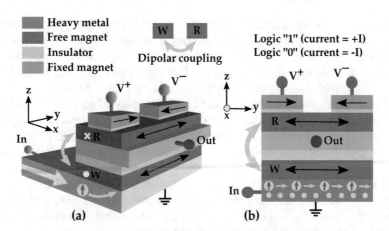

Fig. 2.4 (**a**) Structure and construction of the GSHE switch, adopted from [Ran+17]. (**b**) Cross-section of the GSHE switch

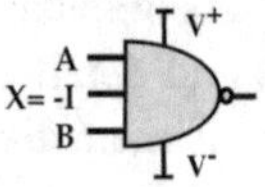

NAND   NOR

A	B	X	Σ In	Out		A	B	X	Σ In	Out
-I	-I	-I	-3I	+I		-I	-I	+I	-I	+I
-I	+I	-I	-I	+I		-I	+I	+I	+I	-I
+I	-I	-I	-I	+I		+I	-I	+I	+I	-I
+I	+I	-I	+I	-I		+I	+I	+I	+3I	-I

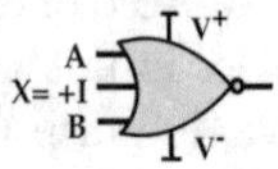

Fig. 2.5 Truth tables for NAND and NOR functionalities, with inputs A and B (X is a control signal). Here, logic 1/0 is represented by an output current +I/-I

the read-out voltage applied to the low-resistance path (i.e., either V^+ or V^-), the output current either flows inward or outward, representing the logic encoding of the GSHE switch operation.

This basic GSHE device can readily implement a BUF or INV gate (buffer or inverter operations). To realize more complex multi-input logic gates, a tie-breaking control signal X with a fixed amplitude and polarity is applied in addition to the primary input signals at the input terminal. Note the input of the GSHE switch (or, more generally, any spin–orbit torque-driven magnetic switch) is additive in nature. The polarities of the control signal and the MTJ voltage polarities are used to permute between different Boolean operations (Fig. 2.5). See also Fig. 2.9 for all 16 possible Boolean gates.

The GSHE switch is a noisy polymorphic device, whose probability for output correctness depends on the input spin current's amplitude and duration, i.e., the outputs generated by the previous logic stages. However, for static camouflaging the spin pulse amplitude and duration are set for deterministic switching.

2.4.1   Characterization and Comparison of the GSHE Switch

To quantify the overheads due to GSHE-based static camouflaging, it is important to evaluate the area, power, and delay metrics of a single device first. Patnaik et al. construct the layout of their GSHE switch (Fig. 2.6) based on the design rules for beyond-CMOS devices [NY13], in units of maximum misalignment length λ. The area of the GSHE switch is accordingly estimated to be $0.0016 \, \mu m^2$.

The material parameters for the GSHE switch considered are given in Table 2.1. A spin current (I_S) of at least $20 \, \mu A$ is required for deterministic computing, whereas sub-critical currents are sufficient for probabilistic computing [Ran+17].

The performance of the switch is determined by nanomagnetic dynamics, simulated using the stochastic Landau–Lifshitz–Gilbert–Slonczewski equation [dAq+06]. Delay distributions for the nanomagnet considered are illustrated in Fig. 2.7. The mean propagation delay for deterministic computing is obtained as 1.55 ns, for $I_S = 20 \, \mu A$. This delay is then used to construct a behavioral Verilog model to obtain the transient responses shown in Fig. 2.8.

The power dissipation for the read-out phase is derived according to the equivalent circuit shown in Fig. 2.6 (inset). Using the following equations and

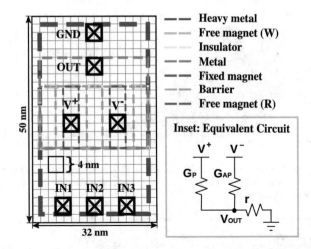

Fig. 2.6 The layout of the GSHE switch constructed according to the design rules for beyond-CMOS devices formulated in [NY13]. Inset shows the equivalent circuit of the GSHE switch, derived from [DSB12]. The power dissipation in the equivalent circuit is dictated by the resistance r of the heavy metal as well as the conductances of the anti-parallel, high-resistance path (G_{AP}) and the parallel, low-resistance path (G_P) composed of the fixed ferromagnets

Table 2.1 Material parameters of the GSHE switch

Parameters	Value
Volume of nanomagnets (NM)	$(28 \times 15 \times 2)\,\mathrm{nm}^3$ [Ran+17]
Saturation magnetization M_S of NM	10^6 A/m (W-NM) [Ran+17]
	5×10^5 A/m (R-NM) [Ran+17]
Uniaxial energy density K_u of NM	2.5×10^4 J/m^3 (W-NM) [Ran+17]
	5×10^3 J/m^3 (R-NM) [Ran+17]
Spin current I_S, determ. switching	$20\,\mu$A [Ran+17]
Resistance area product RAP	$1\ \Omega\mu\mathrm{m}^2$ [Mae+11]
Tunneling magnetoresistance TMR	170% [Mae+11]
Parallel conductance G_P	420 μS
Anti-parallel conductance G_{AP}	155.6 μS
Resistivity of heavy metal (HM) ρ	$5.6 \times 10^{-7}\Omega$-m
Spin Hall angle θ_{SH} of HM	0.4 [HX15]
Thickness t_{HM} of HM	1 nm
Internal gain β of HM	$0.4 \times (15\,nm/1\,nm)$
$\beta = \theta_{SH} \times (w_{NM}/t_{HM})$	= 6
Resistance r of HM	≈ 1 kΩ

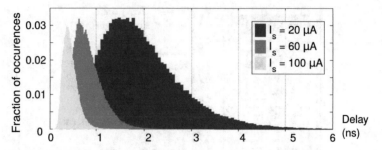

Fig. 2.7 Delay distributions for the GSHE switch at various spin currents (I_S), obtained from 100,000 simulations each. The spread and mean delay diminish with increasing I_S, however, at the cost of higher power dissipation

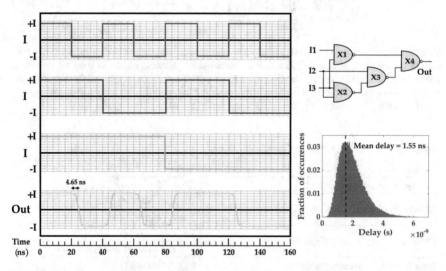

Fig. 2.8 Transient response for all input patterns applied for an example circuit (right top). The critical path comprises three GSHE gates, which each exhibit a mean delay of 1.55 ns (right bottom); hence, the overall delay is 4.65 ns

the parameters listed in Table 2.1, the power dissipation of the GSHE switch for deterministic computing, including leakage, is derived as 0.2125 μW.

$$P = \frac{V_{OUT}^2}{r} + (V_{SUP} - V_{OUT})^2 G_P + (V_{OUT} + V_{SUP})^2 G_{AP} \tag{2.1}$$

$$V_{SUP} = |V^{+/-}| = \left(\frac{I_S}{\beta}\right)\left(\frac{1 + r(G_P + G_{AP})}{G_P - G_{AP}}\right) \tag{2.2}$$

$$V_{OUT} = \frac{I_S\, r}{\beta} \tag{2.3}$$

Table 2.2 Comparison of selected emerging device-based primitives

Device	Functions	Energy	Power	Delay
SiNWFET [Bi+16]	NAND/NOR	0.05–0.1 fJ	1.13–1.77 μW	42–56 ps
ASL [AYF17]	NAND/NOR/AND/OR	0.58 pJ	351.52 μW	1.65 ns
ASL [AYF17]	XOR/XNOR	1.16 pJ	351.52 μW	3.3 ns
ASL [AYF17]	INV/BUF	0.13 pJ	342.11 μW	0.38 ns
DWM [HZ16]	AND/OR	67.72 fJ	60.46 μW	1.12 ns
DWM [Par+17]	NAND/NOR/XOR/ XNOR/AND/OR/INV	N/A	N/A	N/A
GSHE [Zha+15]	AND/OR/NAND/NOR	N/A	N/A	N/A
STT [Win+16]	NAND/NOR/XOR/ XNOR/AND/OR	N/A	N/A	N/A
GSHE (intrinsic)	**All 16**	**0.33 fJ**	**0.2125 μW**	**1.55 ns**

$$\frac{G_P}{G_{AP}} = 1 + TMR; \quad G_P = \frac{A(\text{nanomagnets})}{RAP}. \tag{2.4}$$

Table 2.2 presents a comparison of GSHE switch against other emerging device-based static camouflaging implementations. The higher delay of the GSHE switch is not prohibitive when used in conjunction with delay-aware logic synthesis and gate placement. As for security in terms of obfuscation, the number of possible functions is the relevant metric. Here, the GSHE switch significantly outperforms other devices.

The GSHE switch is leveraged for a versatile and effective security primitive—all 16 possible Boolean functions can be cloaked within a single device (Fig. 2.9). In other words, employing this primitive instead of regular gates can hinder RE attacks of the chip's design IP, without the need for the designer to alter the underlying netlist. For example, to realize NAND/NOR using this primitive, three charge currents are fed into the bottom layer of the GSHE switch at once (Figs. 2.5 and 2.9): two currents represent the logic signals A and B, and the third current (X) acts as the "tie-breaking" control input. For some functions, the logic signals have to be provided as MTJ voltages, not as charge currents. To transduce voltage into charge currents (as well change their polarities), magnetoelectric (ME) transducers are used [Man+19, Ira+17]. These transducers, placed in the interconnects, are capable of charge current conversion (i.e., +I to −I) and voltage to charge current conversion (i.e., high/low voltages to ±I) and vice versa, with relatively low overheads.

Note that three wires must be used for the GSHE input terminals (Fig. 2.6). This is to render the primitive indistinguishable for imaging-based RE by malicious end-users, irrespective of the actual functionality. Several functionalities leave some of those wires unassigned (Fig. 2.9), which can be implemented as non-conductive dummy interconnects [Pat+17, CCW18], especially if only static camouflaging is considered.

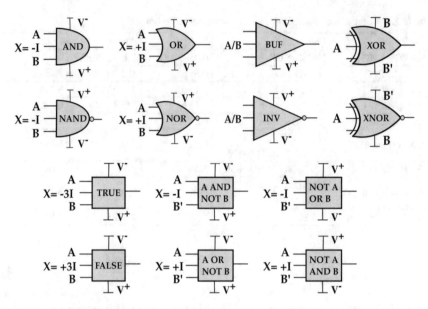

Fig. 2.9 All 16 possible Boolean functionalities for two inputs, A and B, implemented using the proposed primitive. If required, X serves as control signal, not as regular input. Note that BUF and INV capture two functionalities each; hence, only 14 device instances are illustrated

To hinder fab-based adversaries, Patnaik et al. outline two promising options for secure implementation, either (1) leverage split manufacturing [McC11] or (2) provision for a tamper-proof memory. For option (1), the wires for the control inputs and the ferromagnet terminals are to remain protected from the untrusted FEOL fab. Hence, these wires have to be routed at least partially through the BEOL, which must be manufactured by a separate, trusted fab. For option (2), the tamper-proof memory holds a secret key that defines (using some additional circuitry) the correct assignment of control inputs and voltages for all devices. The key must be loaded (by the IP holder) into the memory only after fabrication.

For option (1), fab-based adversaries may employ proximity attacks to recover the withheld BEOL routing from various physical-design hints present in the FEOL [RSK13, Wan+16b]. However, the correct connection rate (which quantifies the BEOL recovery) tends to decrease for larger obfuscation scales. Further, for practically relevant camouflaging scales, the attack by Wang et al. [Wan+16b] incurs excessive runtime. For example, for 30% camouflaging of the benchmark *b21*, when split at M6, the attack cannot resolve within 48 h, resulting in a time-out. Overall, while some BEOL wires can be correctly inferred, proximity attacks are limited, especially when split at lower layers [Wan+16b, Pat+18f, MSD16, Sen+17, Pat+18e].

Both options (1) and (2) represent a notable advancement over prior work related to camouflaging, where the IP holder must trust the fab because of the circuit-level protection mechanism. In the remainder of this chapter, the focus will be on the threats imposed by malicious end-users.

2.4.2 Security Analysis

This section elaborates on the security of the GSHE static camouflaging primitive against various end-user attacks. Most notably, a comprehensive study for analytical SAT attacks is presented, which benchmarks the GSHE static camouflaging against prior art.

2.4.2.1 Threat Model

The malicious end-user is interested in resolving the underlying IP implemented by the obfuscated chip. It is assumed that the attacker possesses the know-how and has access to equipment for imaging-based RE. However, the attacker does not have access to advanced capabilities for invasive read-out attacks, to resolve the voltage and current assignments for individual GSHE gates at runtime.

The attacker procures multiple copies of the chip from open market. They use one for RE, which includes de-packaging, de-layering, imaging of individual layers, stitching of these images, and final netlist extraction [Qua+16b]. The second one is employed as an oracle to obtain input–output (I/O) patterns. These patterns are then utilized for SAT-based attacks.

2.4.2.2 Reverse Engineering Attack Scenarios

1. **Layout identification and read-out attacks:** Since the physical layout of the GSHE static camouflaging primitive is uniform, it remains indistinguishable for optical imaging-based RE. It was also shown that dummy interconnects can become difficult to resolve during RE, as long as suitable materials such as Mg and MgO are used [Hwa+12, CCW18]. A more sophisticated attacker might, however, leverage electron microscopy for identification and read-out attacks. For example, Courbon et al. [CSW16] used scanning electron microscopy, in passive voltage contrast mode, to read out memories in secure chips. However, such attacks would prove significantly more difficult on the GSHE primitive owing to the minuscule dimensions of the GSHE switch, which diminishes spatial resolution for EM-based analysis [CSW16].

2. **Magnetic- and Temperature-driven attacks:** Ghosh et al. [Gho16] explore attacks on spintronic memory devices using external magnetic fields and malicious temperature curves. As for the GSHE switch, note that it is tailored for robust magnetic coupling (between the W and R nanomagnets) [Ran+17], and this coupling would naturally be disturbed by any external magnetic fields.

Hence, an attacker leveraging a magnetic probe may induce stuck-at-faults. However, such malicious incursions are hardly controllable due to multiple factors. First, the extremely small size of the GSHE switch would require very large magnetic fields for the probe. Second, the attacker must know in advance the state of the nanomagnets, the orientation of the fixed magnets, and also the voltage polarities for the MTJ setup. Temperature-driven attacks will impact the retention time of the GSHE switch. The resulting disturbances, however, are stochastic due to the inherent thermal noise in the nanomagnets; fault attacks are accordingly challenging as well. As a result, subsequent sensitization attacks to resolve the obfuscated IP, as proposed in [Raj+13a], will be practically unfeasible.

2.4.2.3 Study on Large-Scale IP Protection Against SAT-Based Attacks

1. **Experimental setup:** The GSHE static camouflaging primitive is modeled as outlined in [Yu+17]. More specifically, the logical inputs a and b are fed in parallel into all 16 possible Boolean gates, and the outputs of those gates are connected to a 16-to-1 MUX with four select/key bits. As for other prior art with less possible functionalities, a smaller MUX with less key bits may suffice. For a fair evaluation, the same set of gates are protected, i.e., gates are randomly selected once for each benchmark, memorized, and then the same selection is reapplied across all techniques. All the static camouflaging techniques are evaluated against powerful state-of-the-art SAT attacks [SRM15, Sub17], with time-out set to 48 h. The experiments are conducted on various benchmarks suites like *ISCAS-85*, *MCNC*, *ITC-99*, *EPFL suite* [Ama15], and the industrial *IBM superblue suite* [Vis+11].
2. **Insights on SAT-based attacks:** Table 2.3 highlights the runtimes incurred by the seminal attack of Subramanyan et al. [SRM15, Sub17]. While there are further metrics such as the number of clauses, attack iterations, or number of remaining feasible assignments [Yu+17], runtime is a straightforward yet essential indicator—either an attack succeeds within the allocated time or not. It is observed that for the same number of gates protected, the more functions a primitive can cloak, the more resilient it becomes in practice. More importantly, the runtimes required for decamouflaging, if possible at all, tend to scale exponentially with the percentage of gates being camouflaged. The GSHE static camouflaging primitive induces by far the highest computational effort across all benchmarks. When over 30% of the circuit is camouflaged using the GSHE primitive, none of the benchmarks could be resolved within 48 h. Further analysis on full-chip protection using the GSHE primitive confirms this superior resilience.
3. **Delay optimization:** Having investigated and benchmarked the SAT-resilience of GSHE-based static camouflaging, the final step in designing the primitive is to reconcile the considerable delay of the GSHE switch with the need to

Table 2.3 Runtime for SAT attacks [SRM15, Sub17], on selected designs, in seconds (Time-Out "t-o" is 48 h, i.e., 172,800 s). Here, * refers to the number of cloaked functions

Benchmark	[Raj+13a] (3)*	[Nir+16, Win+16] (6)*	[Bi+16] (4)*	[Zha16] (2)*	[Zha+15, AYF17] (4)*	[Par+17] (8)*	[Pat+18b] (16)*
10% IP protection							
aes_core	610	4710	890	132	536	6229	25,890
b14	2078	20,603	11,465	6884	17,634	27,438	60,306
b21	7813	162,324	45,465	3977	24,035	t-o	t-o
c7552	37	210	74	12	66	371	2,289
ex1010	62	215	82	12	73	295	922
log2	t-o	t-o	t-o	t-o	t-o	t-o	t-o
pci_bridge32	1119	t-o	9011	1325	2690	t-o	t-o
20% IP protection							
aes_core	4319	41,844	11,306	407	9432	t-o	t-o
b14	56,155	t-o	64,145	8426	t-o	t-o	t-o
b21	t-o	t-o	t-o	t-o	t-o	t-o	t-o
c7552	169	14,575	1153	110	1327	172,548	t-o
ex1010	171	1047	274	38	250	1310	4701
log2	t-o	t-o	t-o	t-o	t-o	t-o	t-o
pci_bridge32	54,577	t-o	t-o	t-o	t-o	t-o	t-o
30% IP protection							
aes_core	17,148	t-o	31,601	2020	26,498	t-o	t-o
b14	56,787	t-o	t-o	38,495	t-o	t-o	t-o
b21	t-o	t-o	t-o	t-o	t-o	t-o	t-o
c7552	1786	t-o	t-o	766	t-o	t-o	t-o
ex1010	448	4357	938	87	719	11,736	24,727
log2	t-o	t-o	t-o	t-o	t-o	t-o	t-o
pci_bridge32	t-o	t-o	t-o	t-o	t-o	t-o	t-o
40–100% IP protection							
aes_core t-o	t-o	t-o	8206	t-o	t-o	t-o	
b14	t-o	t-o	t-o	t-o	t-o	t-o	t-o
b21	t-o	t-o	t-o	t-o	t-o	t-o	t-o
c7552	t-o	t-o	t-o	41,721	t-o	t-o	t-o
ex1010	1703	t-o	129,290	169–7073	1950	t-o	t-o
log2	t-o	t-o	t-o	t-o	t-o	t-o	t-o
pci_bridge32	t-o	t-o	t-o	t-o	t-o	t-o	t-o

camouflage large parts of the target circuit. This can be done by implementing a delay-aware placement strategy, as outlined next. Large-scale circuits typically exhibit biased distributions of delay paths, with most paths having short delays but few paths having dominant critical delays (Fig. 2.10). If the CMOS gates in the non-critical paths are replaced with GSHE-based camouflaged gates, then the excessive delay overheads can be curtailed. On an average, 5–15% of all gates

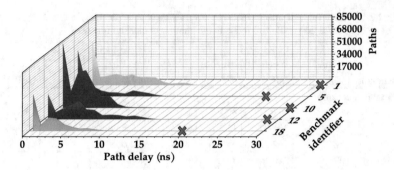

Fig. 2.10 Path delay distributions in large-scale IBM Superblue benchmarks

can be camouflaged this way. Conducting SAT attacks [SRM15, Sub17] on those
protected designs yields no results within 240 h.

2.5   Closing Remarks

In this chapter, we investigated how static camouflaging primitives are constructed
with reconfigurable emerging devices. With advances in physics and materials
engineering, new devices with even better camouflaging properties are expected
in the near future. This chapter equips prospective security researchers with the
necessary know-how to tap into the potential of these devices and formulate novel
static camouflaging techniques. With the insights gained from this chapter, the
readers can devise new primitives from scratch, from conception and selection of
the appropriate emerging technology, to implementing the camouflaged cells and
performing security analysis.

References

[Ama15] L. Amarù, *Majority-Inverter Graph (MIG) Benchmark Suite* (2015). http://lsi.epfl.ch/
MIG
[Aug+11] C. Augustine et al., Low-power functionality enhanced computation architecture
using spin-based devices, in *2011 IEEE/ACM International Symposium on Nanoscale
Architectures* (IEEE, Piscataway, 2011), pp. 129–136
[AYF17] Q. Alasad, J. Yuan, D. Fan, Leveraging all-spin logic to improve hardware security, in
Proceedings of the on Great Lakes Symposium on VLSI 2017 (2017), pp. 491–494
[Beh+10] B. Behin-Aein et al., Proposal for an all-spin logic device with built-in memory. Nat.
Nanotechnol. **5**(4), 266–270 (2010)
[Bi+14] Y. Bi et al., Leveraging emerging technology for hardware security-case study on
Silicon Nanowire FETs and Graphene SymFETs, in *2014 IEEE 23rd Asian Test
Symposium* (IEEE, Piscataway, 2014), pp. 342–347

[Bi+16] Y. Bi et al., Emerging technology-based design of primitives for hardware security. J. Emerg. Tech. Comput. Syst. **13**(1), 3:1–3:19 (2016). ISSN: 1550-4832. https://doi.org/10.1145/2816818

[CCW18] S. Chen, J. Chen, L. Wang, A chip-level anti-reverse engineering technique, in J. Emerg. Tech. Comput. Syst. **14**(2), 29:1–29:20 (2018). ISSN: 1550-4832. https://doi.org/10.1145/3173462

[CSW16] F. Courbon, S. Skorobogatov, C. Woods, Direct charge measurement in floating gate transistors of flash EEPROM using scanning electron microscopy, in *Proc. Int. Symp. Test. Failure Analys.* (2016), pp. 1–9. https://pdfs.semanticscholar.org/992a/20c0a8bb71642fc44fa65f053b3524113b99.pdf

[dAq+06] M. d'Aquino et al., Midpoint numerical technique for stochastic Landau-Lifshitz-Gilbert-dynamics. J. Appl. Phys. **99**(8), 08B905 (2006)

[DSB12] S. Datta, S. Salahuddin, B. Behin-Aein, Non-volatile spin switch for Boolean and non-Boolean logic. Appl. Phys. Lett. **101**(25), 252411 (2012)

[EGT15] M. El Massad, S. Garg, M.V. Tripunitara, Integrated circuit (IC) decamouflaging: reverse engineering camouflaged ICs within minutes, in *NDSS* (2015), pp. 1–14

[Gho16] S. Ghosh, Spintronics and security: Prospects, vulnerabilities, attack models, and preventions. Proceedings of the IEEE **104**(10), 1864–1893 (2016)

[Hir99] J.E. Hirsch, Spin Hall effect. Phys. Rev. Lett. **83**(9), 1834 (1999)

[Hwa+12] S.-W. Hwang et al., A physically transient form of silicon electronics. Science **337**(6102), 1640–1644 (2012). ISSN: 0036-8075. https://doi.org/10.1126/science.1226325. eprint: http://science.sciencemag.org/content/337/6102/1640.full.pdf

[HX15] Q. Hao, G. Xiao, Giant spin Hall effect and switching induced by spin-transfer torque in a $W/Co_{40}Fe_{40}B_{20}/MgO$ structure with perpendicular magnetic anisotropy. Phys. Rev. Appl. **3**(3), 034009 (2015)

[HZ16] K. Huang, R. Zhao, Magnetic domain-wall racetrack memory-based nonvolatile logic for low-power computing and fast run-time-reconfiguration. Trans. VLSI Syst. **24**(9), 2861–2872 (2016)

[Ira+17] R. Mousavi Iraei et al., Electrical-spin transduction for CMOS-spintronic interface and long-range interconnects. IEEE J. Explor. Solid-State Comput. Dev. Circ. **3**, 47–55 (2017). ISSN: 2329-9231. https://doi.org/10.1109/JXCDC.2017.2706671

[Kan+15] N. Kani et al., A model study of an error-free magnetization reversal through dipolar coupling in a two-magnet system. IEEE Trans. Magnet. **52**(2), 1–12 (2015)

[Mae+11] H. Maehara et al., Tunnel magnetoresistance above 170% and resistance–area product of $1\Omega(\mu m)^2$ attained by in situ annealing of ultra-thin MgO tunnel barrier. Appl. Phys. Exp. **4**(3), 033002 (2011)

[Man+19] S. Manipatruni et al., Scalable energy-efficient magnetoelectric spin–orbit logic. Nature **565**(7737), 35–42 (2019)

[McC11] C. McCants, Trusted Integrated Chips (TIC). Tech. rep. Intelligence Advanced Research Projects Activity (IARPA), 2011

[MSD16] J. Magaña, D. Shi, A. Davoodi, Are proximity attacks a threat to the security of split manufacturing of integrated circuits?, in *Proc. Int. Conf. Comp.-Aided Des.* Austin, Texas (2016), pp. 90:1–90:7. ISBN: 978-1-4503-4466-1. https://doi.org/10.1145/2966986.2967006

[Nir+16] I.R. Nirmala et al., A novel threshold voltage defined switch for circuit camouflaging, in *Proc. Europe Test. Symp.* (2016), pp. 1–2. https://doi.org/10.1109/ETS.2016.7519286

[NY13] D.E. Nikonov, I.A. Young, Overview of beyond-CMOS devices and a uniform methodology for their benchmarking. Proc. IEEE **101**(12), 2498–2533 (2013)

[Par+17] F. Parveen et al., Hybrid polymorphic logic gate with 5-terminal magnetic domain wall motion device, in *Proc. Comp. Soc. Symp. VLSI* (2017), pp. 152–157. https://doi.org/10.1109/ISVLSI.2017.35

[Pat+17] S. Patnaik et al., Obfuscating the Interconnects: Low-Cost and Resilient Full-Chip Layout Camouflaging, in *Proc. Int. Conf. Comp.-Aided Des.* (2017), pp. 41–48. https://doi.org/10.1109/ICCAD.2017.8203758

[Pat+18b] S. Patnaik et al., Advancing hardware security using polymorphic and stochastic spin-Hall effect devices, in *2018 Design, Automation & Test in Europe Conference & Exhibition (DATE)* (IEEE, Piscataway, 2018), pp. 97–102

[Pat+18e] S. Patnaik et al., Concerted wire lifting: enabling secure and cost-effective split manufacturing, in *Proc. Asia South Pac. Des. Autom. Conf.* (2018), pp. 251–258. https://doi.org/10.1109/ASPDAC.2018.8297314

[Pat+18f] S. Patnaik et al., Raise your game for split manufacturing: restoring the true functionality through BEOL, in *Proc. Des. Autom. Conf.* (2018), pp. 140:1–140:6. https://doi.org/10.1145/3195970.3196100

[Qua+16b] S.E. Quadir et al., A survey on chip to system reverse engineering. J. Emerg. Tech. Comput. Syst. **13**(1), 6:1–6:34 (2016). ISSN: 1550-4832. https://doi.org/10.1145/2755563

[Raj+13a] J. Rajendran et al., Security analysis of integrated circuit camouflaging, in *Proc. Comp. Comm. Sec.* Berlin, Germany (2013), pp. 709–720. ISBN: 978-1-4503-2477-9. https://doi.org/10.1145/2508859.2516656

[Raj+13b] J. Rajendran et al., Security analysis of integrated circuit camouflaging, in *Proceedings of the 2013 ACM SIGSAC Conference on Computer & Communications Security* (2013), pp. 709–720

[Ran+17] N. Rangarajan et al., Energy-efficient computing with probabilistic magnetic bits–performance modeling and comparison against probabilistic CMOS logic. IEEE Trans. Magnetics **53**(11), 1–10 (2017)

[RSK13] J. Rajendran, O. Sinanoglu, R. Karri, Is split manufacturing secure?, in *Proc. Des. Autom. Test Europe* (2013), pp. 1259–1264. https://doi.org/10.7873/DATE.2013.261

[Sen+17] A. Sengupta et al., Rethinking split manufacturing: an information-theoretic approach with secure layout techniques, in *Proc. Int. Conf. Comp.-Aided Des.* (2017), pp. 329–336. https://doi.org/10.1109/ICCAD.2017.8203796. Revised version available at https://arxiv.org/abs/1710.02026

[SRM15] P. Subramanyan, S. Ray, S. Malik, Evaluating the security of logic encryption algorithms, in *Proc. Int. Symp. Hardw.-Orient. Sec. Trust* (2015), pp. 137–143. https://doi.org/10.1109/HST.2015.7140252

[Sub17] P. Subramanyan. *Evaluating the Security of Logic Encryption Algorithms* (2017). https://bitbucket.org/spramod/host15-logic-encryption

[Vis+11] N. Viswanathan et al., The ISPD-2011 routability-driven placement contest and benchmark suite, in *Proc. Int. Symp. Phys. Des.* (2011), pp. 141–146

[Wal+21] A. Wali et al., Satisfiability attack-resistant camouflaged two-dimensional heterostructure devices. ACS Nano **15**(2), 3453–3467 (2021)

[Wan+16b] Y. Wang et al., The cat and mouse in split manufacturing, in *Proc. Des. Autom. Conf.* Austin, Texas (2016), pp. 165:1–165:6. ISBN: 978-1-4503-4236-0. https://doi.org/10.1145/2897937.2898104

[Win+16] T. Winograd et al., Hybrid STT-CMOS designs for reverse-engineering prevention, in *Proc. Des. Autom. Conf.* (2016), pp. 88–93

[Yu+17] C. Yu et al., Incremental SAT-based reverse engineering of camouflaged logic circuits. Trans. Comput. Aided Des. Integr. Circ. Syst. **36**(10), 1647–1659 (2017). ISSN: 0278-0070. https://doi.org/10.1109/TCAD.2017.2652220

[Zha+15] Y. Zhang et al., Giant spin Hall effect (GSHE) logic design for low power application, in *Proc. Des. Autom. Test Europe* (2015), pp. 1000–1005

[Zha16] J. Zhang, A practical logic obfuscation technique for hardware security. Trans. VLSI Syst. **24**(3), 1193–1197 (2016). ISSN: 1063-8210. https://doi.org/10.1109/TVLSI.2015.2437996

Chapter 3
Runtime Polymorphism for Dynamic Camouflaging

3.1 Chapter Introduction

Polymorphic electronics, introduced by Stoica et al. [SZK01], can potentially revolutionize the way circuit designers protect their intellectual property (IP), via dynamic camouflaging. The ability to dynamically morph the field-deployed target circuit, on-the-fly, can prove invaluable in impeding malicious attacks, especially in the Internet-of-Things (IoT) era. The inherent notion of obfuscating the true functionality of a circuit by continuously morphing it through a series of incorrect functional states, however, introduces an element of error in the computation. This limits the application of dynamic camouflaging to error-tolerant circuits from the neural computation and image processing domains. However, with such approximate neural engines becoming ubiquitous in current energy-efficient systems [Moo+16, Sar+18], protecting their design IP becomes the prerogative of circuit designers.

Typically, CMOS-based polymorphic circuits have relied on the superposition of multiple functionalities through polymorphic logic synthesis, wherein the functional morphing is effected through a change in the temperature, or the supply voltage, or a control signal [SZK01]. This approach, however, is constrained by the limited number of functional states the synthesized circuit can attain. Other implementations involving configurable logic blocks (CLB) and look-up tables (LUT), which can realize far more functionalities, but are too cumbersome to achieve scalability. In this regard, the merit of emerging polymorphic devices is evident from the fact that they can achieve multiple functionalities at a fraction of circuit cost, in terms of area and power.

In this chapter we expound on the application of emerging devices in dynamic camouflaging primitives, by first examining the fundamental concepts and considerations that a designer must keep in mind while trying to fashion a new primitive. This is followed by a brief review of various emerging device candidates, suitable for dynamic camouflaging schemes, and the factors that make them viable. Finally, we delve into an in-depth case study on an existing dynamic camouflaging work, to educate the reader about the various aspects of designing such a primitive.

3.2 Concepts for Dynamic Camouflaging Using Polymorphic Devices

Dynamic reconfiguration in the field is a prerequisite for any emerging device to be considered suitable for constructing a dynamic camouflaging primitive. The key considerations for selecting such a device are as follows:

1. As with static camouflaging, the selected device must be **multi-functional**, either by virtue of its **ambipolarity** or its ability to implement **Majority (MAJ) logic**, with **dummy contacts** to render it **optically indistinguishable** at the layout level. Customarily, dynamic camouflaging would be used to protect only certain parts of a System-on-Chip (SoC), with the other chiplets being fabricated with conventional CMOS technology. Hence, the selected emerging device should be compatible with the existing CMOS processes, to enable hybrid CMOS designs.
2. The device must possess **post-fabrication reconfigurability**, wherein it can be dynamically morphed between various functionalities after deployment. Further, this morphing should be possible for an arbitrary number of times. This is crucial for the end-to-end supply chain protection aspect, since it allows the designer to secure the IP against an untrusted foundry and test facility, in addition to an untrusted end-user. Detailed analyses on these scenarios will be presented in the case study (Sect. 3.4).
3. The device can be programmed to switch between its different functional modes according to a pre-determined control sequence or arbitrarily. In contrast to static camouflaging, the device is continuously morphing in this case. Hence, the **reconfiguration time and energy** of the device are crucial in planning the power and performance budget of the scheme. Ideally, the device should be able to switch functionalities and achieve steady-state operation quickly after reconfiguration, and with minimum side-channel leakage. This allows for minimal overheads, ensuring that the scheme can be used at a larger scale in the target circuit. Thus, emerging devices with fast internal dynamics on the order of *ns*, or smaller, are preferred.
4. Generally, the more the number of functionalities the device can implement, the better it is in increasing the search complexity for an attacker. However, more functionalities also translate to more incorrect functional states for the device, thus exacerbating the computational error. Hence, there is a delicate trade-off between the accuracy of the protected approximate circuit and the security afforded by the dynamic camouflaging scheme implemented.

3.3   Review of Selected Emerging Technologies and Prior Art

In this section, we briefly review various classes of emerging devices and phenomena, which are promising for implementing a successful dynamic camouflaging scheme. This survey is intended to provide security researchers with insights into how they can select an appropriate polymorphic device for their primitive.

3.3.1   Magneto-Logic Devices

Dery et al. [Der+11] proposed a reconfigurable magneto-logic gate composed of a hybrid graphene–ferromagnet structure, with excellent CMOS compatibility. Apart from the $O(ns)$ speed of operation and high scalability, the device also showcases good spin amplification owing to its graphene channel.

The construction of the magneto-logic device is as shown in Fig. 3.1. The five stacks on top of the graphene channel consist of CoFe/Cu/Py/MgO layers. Each stack has two CoFe/Cu electrodes on an elongated Permalloy (Py) strip, where the CoFe ferromagnets act as the fixed layers and the Py is employed as the free layer. The fixed ferromagnets are oriented opposite to each other, and the all-metallic path (CoFe/Cu/Py/Cu/CoFe) for electric current conduction between these pairs of electrodes on each stack ensures low writing power.

The working of the magneto-logic gate is as follows. First, an input current pulse between the CoFe contacts of each stack (A–D) sets the Py magnetization of that stack through spin-transfer torque (STT) switching. Note that the direction of the current pulse decides the final magnetization orientation of the Py free layer. Here, the side stack (A–D) contacts represent the primary input terminals, whereas the middle contact M is reserved for read-out. Once the free layers are set into their final magnetization state, the tunneling of spins from the Py free

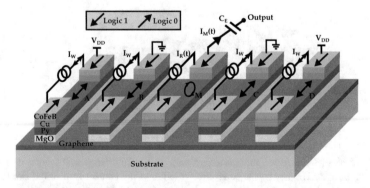

Fig. 3.1 Schematic of the universal magneto-logic gate. Here, A–D are the primary input terminals and M is the read-out terminal

layers into the graphene, via the MgO barriers, results in a non-equilibrium spin accumulation in the graphene. The relative orientations of the Py free layers decide the magnitude of this spin accumulation. The logic is generated by applying a small perturbation to the middle free layer magnetization. The transient current response to this perturbation is dictated by the potential level in the middle contact, which in turn depends on the spin accumulation profile in the graphene. Hence, the primary inputs on the side terminals determine the response and thus the logic output.

The logic implemented by this structure is $A \oplus B + C \oplus D$, which can be extended to implement other Boolean gates by tying the B and C inputs to high/low signals. An OR(A,D) gate can be realized by setting $B = C = 0$, while $B = C = 1$ implements NAND(A,D) operation. A OR NOT D is achieved when $B = 1$ and $C = 0$. Essentially B and C act as control signals here, which can be changed on-the-fly to enable runtime polymorphism. The presence of this runtime polymorphism, coupled with its scalability and CMOS compatibility, makes the magneto-logic device a viable candidate for constructing a dynamic camouflaging primitive.

3.3.2 Domain Wall Devices

The composite multilayer structure demonstrated in [Ang+20] is used to implement a hybrid CMOS-domain wall technology-based polymorphic gate. The domain wall strip (DWS) in the structure is a free ferromagnetic layer composed of CoFeB, which can support the formation of multiple magnetic domains. This strip is used as the common free layer for three Magnetic Tunnel Junction (MTJ) arrangements, as shown in Fig. 3.2. The DWS is terminated by two stabilizing ferromagnets at either end, which are oriented anti-parallel to each other. On applying a potential across the Write terminals (W^+ and W^-) of the device, the stabilizers polarize the input electric current into a spin-polarized current and supply it to the DWS. This spin

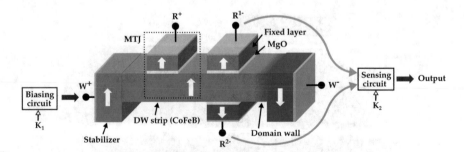

Fig. 3.2 Hybrid polymorphic gate constructed with composite domain wall motion device, and CMOS peripherals. The key inputs control the position of the domain wall in the five-terminal device, as well as the read path for output sensing, thus deciding the logic implemented

current results in the STT-induced motion of the DW (formed between the domains in the DWS) in the direction of the electron flow.

The triple MTJ arrangement on the DWS creates two distinct read paths, one from R^+ to R^{1-} and another from R^+ to R^{2-}. These paths provide different resistance states to be detected by the read current, depending on the free layer magnetization, which in turn depends on the position of the DW inside the strip. The CMOS bias circuitry is responsible for providing an input current bias to position the DW, while the sensing circuit reads out the resistance state, which is a function of the DW position relative to the MTJ read paths. The primary inputs A and B are fed through the bias circuit. The key inputs in the bias and sense circuits provide external control to manipulate the logic, by setting the input bias and selecting the appropriate read path, respectively. Different combinations of input bias (and hence, DW position) and read paths result in NOT/AND/NAND/OR/NOR/XOR/XNOR functions, and the device can morph between these functions during runtime if the key signals are modified. Details of the different key configurations to achieve various logic functionalities can be found in [Ang+20]. Although this hybrid polymorphic gate requires a considerable number of peripheral CMOS transistors, the large number of functions implemented with the same layout makes it a promising contender to implement dynamic camouflaging schemes.

3.3.3 Topological Insulator-Based Devices

Topological insulators (TI) represent a unique material class, which can enable the next generation of high-performance and polymorphic logic devices. This can be attributed to their significantly larger spin Hall conductivity [KUH18], coupled with their extremely small longitudinal dissipation [BB16]. Rakheja et al. leverage these properties to construct their voltage-controlled topological spin switch (VTOPSS) in [RFK19].

The construction of the VTOPSS device is highlighted in Fig. 3.3. An electric field applied across the TI layer results in a transverse spin current due to the colossal spin Hall effect. This spin current imparts a torque to the spins in the magnetic insulator (MI) layer below, via exchange coupling or antidamping mechanisms. The small damping coefficient of the MI allows its magnetization to switch rapidly (sub-ns), in response to the applied spin torque. The information stored in the MI magnetization is then read-out via an MTJ arrangement. The free layer of the MTJ is exchange coupled to the MI and follows its state. The two voltage terminals on the oppositely oriented fixed magnets of the MTJ allow the application of differential voltages (V^+ and V^-) in the read unit. The output voltage polarity follows whichever voltage terminal lies on the low-resistance read path (whose fixed magnet is parallel to the free layer magnetization). Here, the logic is encoded in the polarity of the output voltage. Hence, interchanging the V^+ and V^- terminals in the read unit will negate the logic operation implemented.

Fig. 3.3 Basic structure of the voltage-controlled topological spin switch, implementing copy or negation operations

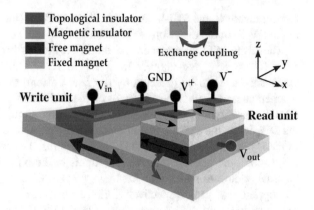

Complex two-input Boolean gates can be realized by including two additional TI input terminals. The primary inputs are provided at two of the terminals, and a tie-breaking control signal is fed into the third one. A negative control signal implements NAND operation, while a positive control signal results in NOR. XOR/XNOR is achieved by supplying one of the operands as an input voltage to the TI and the other one as the read voltage in the MTJ. A common three-input-terminal VTOPSS structure can possibly implement all 16 Boolean functions of two variables, using the same device layout. Note that this common layout would require dummy inputs to mask device functionality.

3.4 Case Study: Dynamic Camouflaging with Magnetoelectric Spin–Orbit (MESO) Devices

Dynamic camouflaging involves obfuscating and switching the device-level functionality post-fabrication, as well as during runtime, thereby hindering various attacks throughout the IC supply chain. This case study demonstrates dynamic camouflaging using polymorphic MESO devices and highlights how security and computational accuracy are two entangled design variables while implementing dynamic camouflaging, especially for error-tolerant applications such as image processing and machine learning. As arbitrary dynamic morphing can be impractical and result in excessive loss of computational accuracy, this study considers dynamic camouflaging based on runtime reconfiguration among functionally equivalent or approximately equivalent circuit structures with the help of polymorphic gates, while maintaining the practicality of such circuits. For such applications, dynamic camouflaging can thwart both exact [SRM15, MGT15] and approximate SAT (*AppSAT*) attacks [Sha+17a]. The resulting end-to-end protection of the IC supply chain is discussed extensively with regard to circumventing the risks associated with untrusted foundries, test facilities, and end-users. The full spectrum of security

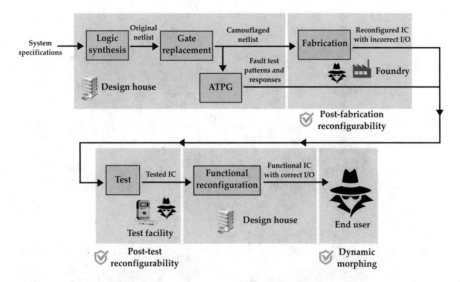

Fig. 3.4 Threat model for dynamic camouflaging-based IP protection. Green and red blocks represent the *trusted* and *untrusted* entities, respectively. The protection schemes—which are all flavors of dynamic camouflaging—employed for each of the untrusted entities are mentioned below the respective red blocks, and indicated by green shields. Gate replacement, which can either be random or through some designer's chosen heuristic, involves the selective replacement of gates in the original netlist with polymorphic MESO gates. After fabrication and testing, the design is sent back to the design house (or some trusted facility) for functional reconfiguration before deployment. ATPG stands for automatic test pattern generation

Table 3.1 IP protection techniques versus untrusted entities in IC supply chain (✓: protection offered, ✗: no protection offered)

Technique	FEOL/BEOL	Test facility	End-user
Logic locking	✓/✓	✓ ([Yas+16a])	✓
Layout camouflaging	✗/✗ (✓/✗ [Pat+20b])	✗	✓
Split manufacturing	✓/✗	✗	✗ (✓ [Pat+18d, Pat+19a])
Dynamic camouflaging	✓	✓	✓

guarantees offered by dynamic camouflaging and the threat model it caters to can be appreciated from Fig. 3.4 and Table 3.1.

3.4.1 MESO Fundamentals

The emerging spin device considered in this study is the MESO device, whose operation is based on the phenomena of magnetoelectric (ME) switching [Lot+04] and inverse spin–orbit effects [DP71]. Proposed by Manipatruni et al. [Man+19], it is capable of implementing different Boolean functions, as shown in the schematics

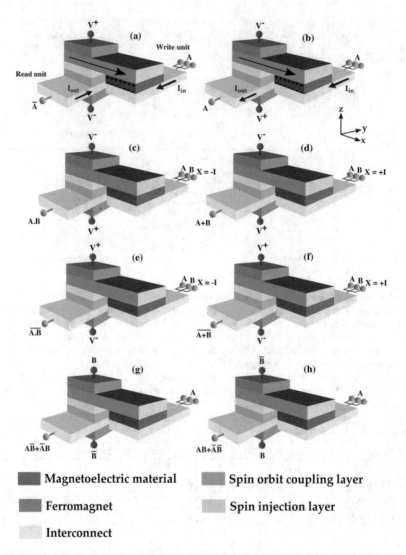

Fig. 3.5 (**a–h**) Implementation of INV, BUF, AND, OR, NAND, NOR, XOR, XNOR with a single MESO device, using different input configurations. Signals *A* and *B* are logic inputs, and *X* is a control input required for some functionalities. Note that INV, BUF, XOR, and XNOR gates have dummy wires/contacts at their input terminals, to make them optically indistinguishable from other implementations

in Fig. 3.5. Here, the inputs/outputs are electric currents, and the logical information is encoded in the direction of the current flow.

During the writing phase, an input electric current flowing in the $\pm\hat{y}$ direction through the non-magnetic interconnect sets up an electric field in the $\pm\hat{z}$ direction within the ME capacitor (red in Fig. 3.5). The resulting ME field switches the magnetization state of the ferromagnet (purple) along the $\pm\hat{x}$ direction. Information is written into the MESO device by transducing the input electric current into the magnetization state of the device. Typical room-temperature multiferroics used for the ME capacitor include $BiFeO_3$ and $LuFeO_3$. The charge accumulation across an ME capacitor in response to an applied electric field is given as $Q_{ME} = A_{ME}(\epsilon_0\epsilon_{mf}E + P_{mf})$, where A_{ME} is the cross-sectional area of the capacitor, $\epsilon_0 = 8.85 \times 10^{-12}$ F/m is the permittivity of free space, ϵ_{mf} is the relative dielectric permittivity of the ME, and P_{mf} is the saturated ferroelectric polarization. For the $BiFeO_3$ capacitor considered in [Man+19], $A_{ME} = 10^{-16}$ m^2, while $\epsilon_{mf} = 54$. The electric field to be applied to the ME capacitor to switch it all-electrically is $E = E_{mf}B_c/B_{mf}$, where $E_{mf} = 1.8 \times 10^6$ V/m refers to the electric switching field, $B_{mf} = 0.03$ T is the exchange bias at switching field, and $B_c = 0.1$ T is the ME switching field.

After the writing process is complete, which takes $\sim$200 ps [Man+19], the supply voltages V^+ and V^- are turned on to initiate the reading phase. In the reading phase, a spin-polarized current is injected into the spin–orbit coupling (SOC) layer, which converts the spin current into electric current at the output node (I_{out}), due to the inverse spin Hall and inverse Rashba–Edelstein effects [SVR14]. These topological effects result in the shifting of the Fermi surface of the high-SOC material in k-space. This shift causes a charge imbalance and hence a charge current in the Fermi surface, in a direction orthogonal to the injected spin density. The magnitude of the charge current transduced by the SOC layer as a result of the applied spin density is given by

$$j_c = \frac{\alpha_R \tau_s}{\hbar} j_s = \lambda_{IREE} j_s, \qquad (3.1)$$

where α_R is the Rashba coefficient, τ_s is the spin relaxation time, and λ_{IREE} ($\sim 1.4 \times 10^{-8}$ m) is the inverse Rashba–Edelstein length of the SOC material [Man+19].

The direction of the output current is determined by the polarity of the supply voltages V^+/V^- ($\pm$100 mV) applied on the nanomagnet, and the final magnetization state of the ferromagnet. For instance, when the ferromagnetic moment is along $+\hat{x}$ and the flow of the injected spin current is along $-\hat{z}$, with spin polarization along $+\hat{x}$, the direction of the charge current generated is along $+\hat{y}$ (Fig. 3.5a). However, when the ferromagnet is reversed to the $-\hat{x}$ direction, with the injected spin current direction unchanged but the spin polarization now along $-\hat{x}$, the output charge current reverses to $-\hat{y}$. The same reversal in the direction of output current can also be achieved by keeping the ferromagnetic moment constant and flipping the voltage polarities V^+/V^-.

The total intrinsic switching time of the MESO device is a combination of the time taken to charge the multiferroic capacitor, τ_{ME}, and the ferroelectric polarization/magnetization reversal time, τ_{mag}. These are given as $\tau_{ME} = 2Q_{ME}/I_{ISOC}$ and $\tau_{mag} = \pi/\gamma B_c$, where I_{ISOC} is the current produced by the spin–orbit effect, and γ is the gyromagnetic ratio of the electron. Evaluating these switching times according to the parameters in [Man+19] yields an intrinsic switching time of $\sim$230 ps. The total switching time of the MESO device is then obtained as ∼258 ps, by adding the interconnect delay of 2.9 ps and the extrinsic peripheral delay of ∼25 ps. For a further, in-depth analysis about the switching and transduction processes in the MESO device, interested readers are referred to [Man+19].

3.4.2 Functional Polymorphism with MESO Gates

The MESO device can implement a buffer (BUF) or an inverter (INV) using the same device (Fig. 3.5a,b), by switching the polarity of the supply voltages. Further, complex gates, such as majority logic, can be realized by leveraging the additive nature of the input signals. As shown in Fig. 3.5c,d, A and B are the signal inputs and X is the tie-breaking control input. The polarity of X decides the functionality of the MESO gate. Here, $X = -I$ results in an AND gate and $X = +I$ performs an OR operation. To implement NAND and NOR gates, the polarities of the supply voltages are flipped (Fig. 3.5e,f). For XOR and XNOR gates, the tie-breaking input X is eliminated, and only one input is provided at the input terminal. The other input signal is encoded in the voltage domain and applied directly at the V^+/V^- terminals (Fig. 3.5g,h).

Illustrative waveforms showing the device operation and functional reconfiguration between AND/OR and NAND/NOR, on flipping the control signal X, are shown in Fig. 3.6. The MESO device with additional peripheral circuitry is shown in Fig. 3.7. The control bits deciding the input and control signals can either be derived from (1) a control block (Fig. 3.8) or (2) a true random number generator (TRNG), if random reconfiguration is required. The latter case is applicable for error-tolerant applications such as image (video) processing, machine learning, etc. Configuring the MESO device via different supply voltages and electric currents allows us to dynamically implement all basic Boolean gates within a single structure, which is the primary requirement for dynamic camouflaging as seen before.

The design house can either provide a fully camouflaged layout composed of only MESO devices or a camouflaged layout where selected CMOS gates are replaced by MESO gates. Spin devices like MESO are generally compatible with CMOS processes in the BEOL, enabling heterogeneous integration [Yog+15]. The proportion of the design camouflaged by a designer depends on the scope of application and impact of camouflaging on PPA overheads. The replacement of logic gates can also be performed in a manner conducive to protecting the critical infrastructure (i.e., design secrets, proprietary IP).

Note that the MESO-based primitive can also be leveraged for static camouflaging, as delineated in Chap. 2. In such a scenario, the peripheral circuitry (Fig. 3.7) dictating the functionality of the MESO device shall be fed with fixed control bits and control signals.

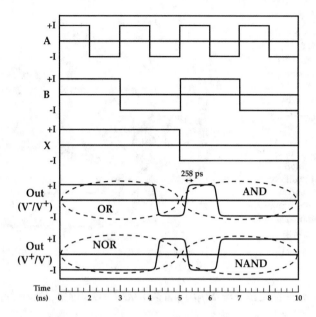

Fig. 3.6 Timing waveforms for MESO-based AND/OR/NAND/NOR gates from behavioral Verilog models, which represent the estimated overall delays of the MESO device along with their intended functionality. The MESO primitive's function morphs based on the value of the voltage terminals and the control signal X. Toggling X allows one to morph between OR $\leftrightarrow$ AND and NOR $\leftrightarrow$ NAND. Morphing between OR and NOR involves setting the top/bottom voltage terminals as V^-/V^+ or V^+/V^-; the converse is true for AND $\leftrightarrow$ NAND morphing. $\sim$258 ps, in the form of the peripheral MUX delay

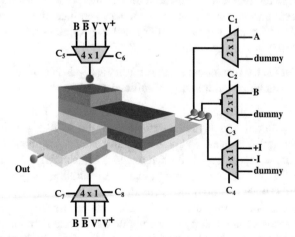

Fig. 3.7 A generic MESO gate with peripheral MUXes, which dictate the input and control signals through control bits C_1–C_8. This generic structure implements any of the Boolean functionalities in Fig. 3.5(a–h) once the appropriate control bits are provided

Fig. 3.8 MESO
adder/subtractor highlighting
the capabilities for functional
reconfiguration. The XOR
and AND gates are
implemented as static MESO
gates and the INV/BUF is a
polymorphic MESO gate
whose function is derived
from control bits (C_1 and C_2)
fed by a simple control block.
A and *B* are the inputs, *S* is
the sum, *D* is the difference,
Ca is the carry, and *Bo* is the
borrow. Note that dummy
contacts are omitted here for
the sake of simplicity

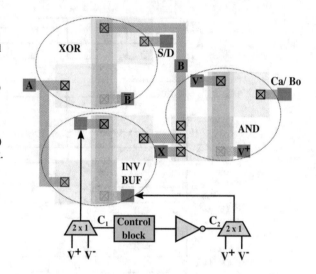

3.4.3 Dynamic Camouflaging Against Untrusted Foundry

While reverse engineering a hybrid MESO–CMOS design, an attacker in the
foundry can readily infer the IP implemented in CMOS, whereas the MESO
gates appear as white-box devices without any fixed functionality. The MESO
implementation of Boolean gates is optically indistinguishable with respect to their
physical layout (Fig. 3.8), which renders optical inspection-guided RE difficult. The
post-fabrication reconfigurability aspect of MESO-based dynamic camouflaging
makes it intuitively resilient to "inside foundry" attacks and hinders the attacker's
effort to infer the exact functionality. A random gate-guessing attack on a circuit
with N number of camouflaged MESO gates has a solution space of 8^N possible
netlists, with only one being correct.

Threat Model The threat model for the security analysis of dynamic camouflaging,
with respect to an untrusted foundry, is outlined as follows:

- A malevolent employee in the foundry has access to the physical design,
  including material and layout parameters of the MESO gates and the chip
  interconnects. While an adversary in a foundry can readily obtain the dimensions
  and material composition of the nanomagnet in each MESO gate and, hence,
  understand its magnetic properties including saturation magnetization, energy
  barrier, and critical ME field for switching, these design details do not leak any
  information about the intended functionality implemented by the gate.
- He/she is aware of the underlying gate selection algorithm, number, and type of
  camouflaged gates but is oblivious to the actual functionality implemented by the
  camouflaged gate.

- For security analysis, we assume that the working chip is not yet available in the open market. Thus, he/she has to apply "inside foundry" attacks that are explained briefly next.

Attack Model Recently proposed attacks like [Mas+17, LO19] can be carried out within the confines of an untrusted foundry. These attacks do not leverage an activated working chip as an oracle, which is in contrast with algorithmic SAT-based attacks [SRM15, MGT15, Sha+17a]. Though these attacks have been primarily tailored toward logic locking (LL), we believe these would readily apply on layout camouflaging (LC) schemes as well, given that any LL problem can be modeled as an LC scheme and vice versa. Besides, for the attacks proposed in [Mas+17, LO19], the basic premise is that an incorrect assignment of key bits involves significant logic redundancies compared to the correct assignment of key bits. The attack in [LO19] determines the likely value of key bits individually by comparing the levels of logic redundancy for each logic value.

The effect of an incorrect assignment of key bits (gates) on logic redundancy is illustrated using a simple example. Consider the circuit shown in Fig. 3.9a, where logic gates U31 and U33 are camouflaged using a NAND/NOR camouflaging primitive, resulting in four combinations for [U31, U33]. The circuits are shown in Fig. 3.9b–d, and they correspond to the scenarios, [U31 = NAND, U33 = NAND], [U31 = NAND, U33 = NOR], and [U31 = NOR, U33 = NOR], respectively. After

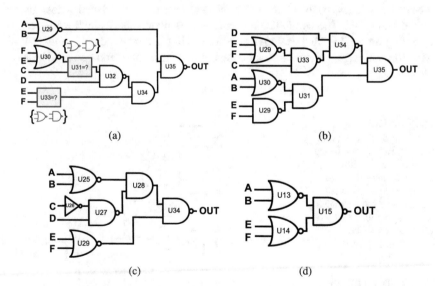

Fig. 3.9 An illustration of an incorrect gate assignment leading to logic redundancy. In (**a**), gates U31 and U33 are camouflaged using a simple NAND/NOR primitive, giving rise to four possible options. Correct assignment of camouflaged gates is shown in blue. Incorrect assignment of gates leads to circuit configurations (**b**), (**c**), and (**d**), respectively. Note the reduction of gates in (**c**) and (**d**) compared to (**a**), while the gate count is identical in (**a**) and (**b**), albeit (**b**) functions differently than (**a**)

re-synthesis, an incorrect combination of gates deciphered by an attacker leads to circuits with fewer gates (Fig. 3.9c and d) when compared to the original circuit. Also note that an attacker might end up with cases like that of Fig. 3.9b, where the total number of gates is the same as the original circuit; however, these circuits differ in functionality.

Performing quantitative experiments, based on the findings quoted in [Mas+17, LO19], the desynthesis attack [Mas+17] can correctly infer 23 (up to 29) and 47 (up to 59) key bits for 32 and 64 key-gates, respectively, while [LO19] report success rate in 25–75% percentile distribution.

MESO Primitive Setup and Modeling In the absence of well defined standard-cell libraries, emerging device-based primitives are often simulated using CMOS MUX models. For this case study, the MESO primitive is modeled as shown in Fig. 3.10. The logical inputs a and b are fed in parallel into all eight possible Boolean functions, and outputs of those gates are connected to an 8-to-1 MUX with three select lines/key bits. For a fair evaluation, the same set of gates are to be camouflaged in the chosen ISCAS-85 benchmarks c5315 and c7552. Gates are selected randomly at the beginning and then memorized. Ten such sets are created for each benchmark. To emulate the attack results from [Mas+17, LO19], the following procedure is employed. A simple script to randomly pick the correct assignment among the camouflaged gates is used to obtain three sets, each corresponding to 50%, 70%, and 90% correctly inferred gates. This procedure is repeated ten times, each for ten different iterations of camouflaged gates, giving 100 unique trials. The camouflaging scheme is implemented using *Python* scripts operating on *Verilog* files. Hamming distance (HD) is computed on *Synopsys VCS* with 100,000 input patterns and functional correctness is ascertained by *Synopsys Formality*.

Fig. 3.10 Modeling of camouflaged circuits using MUXes. Each camouflaged gate is replaced with a corresponding MUX that dictates the functionality based on the value assigned to the select inputs (A1–A6). In this example, the camouflaged cell can implement any one of the eight functions viz. OR, NOR, AND, NAND, INV, BUF, XOR, and XNOR

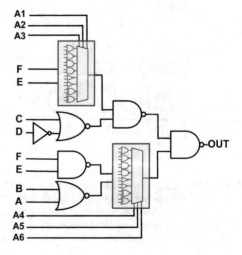

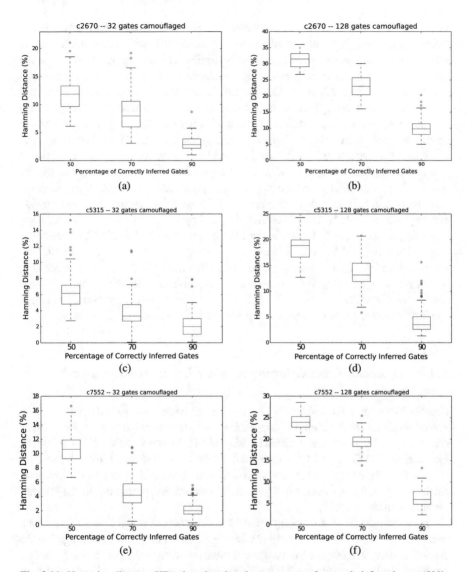

Fig. 3.11 Hamming distance (HD) plotted against the percentage of correctly inferred gates (50%, 70%, and 90% of the total camouflaged gates) of different sets of camouflaged gates, on selected ISCAS-85 benchmarks (**a–b**) c2670, (**c–d**) c5315, and (**e–f**) c7552. Mean HD is proportional to the number of correctly inferred gates among the total number of camouflaged gates. Each box comprises data for 100 trials of random selection of gates to camouflage

Once the percentage of correctly inferred gates for different levels of attack accuracy (50%, 70%, and 90%) is ascertained, the HD between the reconstructed and the golden netlist can be calculated. Sample results are shown as box-plots in Fig. 3.11 for two ISCAS-85 benchmarks c5315 and c7552, for the reader's understanding. It is intuitive to note that, as the percentage of the correctly inferred gates is increased, there is a steady reduction in the HD, which also hints that the reconstructed netlist becomes functionally similar to the original circuit. For the ISCAS-85 benchmark c7552, assuming an attack accuracy of 90%, the mean HD increases from about 2% when 32 gates are camouflaged (29 are inferred correctly) to 5% when 128 gates are camouflaged (115 are inferred correctly). Note that such HD numbers could already suffice for an attacker recovering an approximate version of the original functionality. However, for attacks that can only recover 50–70% of the total camouflaged gates, the HD for the reconstructed circuit is between 6% and 25%, depending on the size and type of the benchmark, the number of gates being camouflaged, and the number of gates correctly inferred. These findings also imply that camouflaging large parts of a design might suffice to thwart "inside foundry" attacks, which can be confirmed by a simple experiment as discussed next. Consider, for example, that for a larger ITC-99 benchmark like b22_C, the designer camouflages 50% of the total logic gates (7228 gates). Assuming that an attacker can identify 90% of these gates correctly, this still leaves 722 gates wrongly inferred, which yields an HD of 43% (across ten random trials). Overall, the property of post-fabrication reconfigurability for the MESO gates allows the design house to change the functionality, enabling superior security through dynamic camouflaging.

3.4.4 Dynamic Camouflaging Against Untrusted Test Facility

Attackers present in the test facility, having access to test patterns and corresponding output responses (generated and supplied by the trusted design house), can jeopardize the security guarantees offered by LL and LC. Modern Automatic Test Pattern Generation (ATPG) algorithms have been designed to maximize the fault coverage (FC) with minimal test pattern count, which directly translates to a lower test cost. Such an approach, however, divulges critical information pertaining to the internal circuit specifics [YSR17].

In the context of VLSI testing principles, detection of a stuck-at-fault involves two principal components, namely (1) fault activation and (2) fault propagation. In fault activation, a faulty node is assigned a value opposite to the fault induced on that particular node. Consider the example shown in Fig. 3.12. Here, the output of gate U4 is *s-a-1* (stuck-at-1). In order to detect this fault, fault activation is achieved by setting this node to logic "0." Next, fault propagation entails propagating the effect of the fault along a sensitization path to one of the primary outputs. To achieve fault propagation (here to O2), the output of U3 must be "1." An input pattern that can detect a fault at a given node by achieving the effects mentioned above is defined as a test pattern. In Fig. 3.12, the input pattern 11001 and the corresponding output response 11 are supplied to the test facility, among others.

Threat Model Apart from outsourcing of chip fabrication, many design companies also outsource the testing phase to off-shore companies such as Amkor, ASE, and

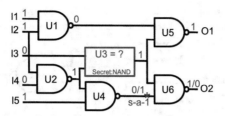

Fig. 3.12 An input pattern that helps in the detection of stuck-at-1 fault at the output of U4. The circuit output "1/0" at O2, which indicates the response for fault-free and faulty circuit are 1 and 0, respectively. The input pattern 11001, along with the expected output response 11, is provided to the test facility for testing manufactured ICs. The test data hints that U3 cannot be NOR. Note that, here the camouflaged gate can function only as NAND/NOR

SPIL [YSR17]. The implications of an untrusted test facility in the supply chain have been explored in the context of LL [Yas+16a] and static camouflaging [YSR17]. This case study, on the other hand, presents an analysis on the efficacy of test data-based attacks [YSR17] for dynamic camouflaging. In this regard, the attacker residing in the test facility is assumed to have access to the following assets:

- Gate-level camouflaged netlist, obtained via RE.
- Knowledge of the test infrastructure, which includes identification of scan chains, compressor, decompressor, etc., on the target chip.
- Test patterns and their corresponding output responses, which have been provided by the design house. He/she also has access to ATPG tools used to generate the test patterns.

Attack Model Yasin et al. [YSR17] proposed *HackTest*, which revealed the true identity of camouflaged gates within minutes by exploiting test data. The attack leverages the fact that the generation of test patterns is typically tuned to obtain the highest possible FC. Hence, given the test stimuli and responses, an attacker can search over the key space (using optimization techniques) to infer the correct assignment of camouflaged gates that maximize the FC. Arguably, such an attack is more powerful than SAT-based attacks [SRM15, MGT15] that require access to a working chip.

The process of ATPG is highly dependent on the internal specifics of the underlying circuit, which include the type and count of gates, the inter-connectivity among these gates, etc. Years of research have yielded powerful algorithms that lower the test pattern count while achieving a high FC. However, these algorithms do not factor in security, and thereby, test patterns become a rich source of information for an opportunistic attacker. Next, working of *HackTest* is explained briefly with a simple example; interested readers are referred to [YSR17] for further details.

Upon performing ATPG for the circuit shown in Fig. 3.13, for the correct assignment of two camouflaged gates (U22 = OR and U28 = AND), eight test patterns are generated by *Synopsys Tetramax*, providing a fault and test coverage of 100%. Camouflaging two gates with two functions each gives rise to four

Fig. 3.13 An example circuit where two gates (U22 and U28) are camouflaged. The camouflaged gates can assume any one of the outlined six 2-input functions. The correct functionality of these camouflaged gates is shown in blue

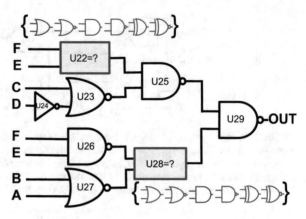

Table 3.2 Fault coverage achieved for different assignments to the camouflaged netlist in Fig. 3.13. Here, U22 and U28 are assumed to implement either AND/OR. The correct assignment is OR and AND, respectively; note that other assignments result in significantly lower fault coverage

U22	U28	Fault coverage (%)
AND	AND	63.33
AND	OR	38.33
OR	AND	100
OR	OR	78.33

possible circuit configurations; Table 3.2 denotes the FC for these configurations. Armed with input patterns and the corresponding output responses, both tailored for the correct assignment, an attacker calculates the FC for all possible circuit configurations. As shown in Table 3.2, maximal FC is observed only for the correct assignment of camouflaged gates. This is because, for static camouflaging, test patterns have to be generated for the correct assignment of camouflaged gates. An attacker can easily use FC to guide his/her attack to identify the correct functionality of camouflaged gates.

MESO Modeling and Setup for HackTest Selected benchmarks from the ISCAS-85 and ITC-99 suites are chosen to run *HackTest*, and the statistics of these benchmarks like the number of logic gates (# Gates), number of faults (# Faults), number of test patterns generated by *Synopsys Tetramax* (# Test patterns), and corresponding FC are shown in Table 3.3. The MESO-based camouflaging primitive and other selected prior art [Raj+13a, Bi+16] are modeled and prepared for *HackTest* as follows. As *HackTest* requires a *BENCH* file format, custom scripts are used to convert *Verilog* files into required formats. For the small-scale ISCAS-85 benchmarks, ten random sets each for camouflaging 32, 64, and 128 gates are prepared. For the large-scale ITC-99 benchmarks, 350 gates are camouflaged. For the sake of uniformity in comparison, the selection of camouflaged gates is random but fixed, i.e., they are common and maintained across all benchmarks for any given

Table 3.3 Statistics of ISCAS-85 and ITC-99 benchmarks used in this case study. All benchmarks achieve 100% test coverage and achieve ~100% fault coverage

Benchmark	# Gates	# Faults	# Test patterns	Fault coverage (%)
c880	273	1764	63	100
c1908	230	1462	80	100
c2670	433	2936	134	100
c3540	814	5472	177	100
c5315	1232	7708	124	100
c7552	1197	7474	167	100
b14_C	4125	24, 668	470	99.99
b15_C	6978	42, 310	812	99.96
b20_C	9226	54, 894	897	99.89
b22_C	14, 457	85, 852	1356	99.95

camouflaging scheme. The attack is implemented using custom *Python* scripts on *Synopsys Tetramax*, with a time-out (t-o) of 24 h.

Impact of Dynamic Camouflaging on HackTest   Since none of the static camouflaging approaches [Raj+13a, Li+16] allow for post-fabrication reconfiguration, their test patterns are generated for the correct assignment of camouflaged gates. Hence, *HackTest* performs extremely well in this scenario. However, MESO-based dynamic camouflaging circumvents this threat by allowing for *post-test configuration*. That is, the fabricated IC can be initially configured with an incorrect I/O mapping and functionality. The "falsely configured" IC and related test data are then sent to the test facility. Accordingly, an attacker will end up with an incorrect IP when mounting *HackTest* on the IC. Note that this is analogous to the idea of *post-test activation* [Yas+16a], which is the adopted strategy for safeguarding against untrusted test facilities in logic locking. After testing is finished, the MESO gates are reconfigured (by the design house or some trusted entity) to reflect the true, intended functionality.

Table 3.4 details the effect of increasing the number of possible functions implemented by the MESO-based primitive, on *HackTest*'s accuracy. It is observed that the attack accuracy reduces (for the same set of camouflaged gates) when the number of functions implemented by the MESO-based primitive is increased. This can be reasoned from the fact that, with an increase in the number of possible functions, the attack has a larger solution space to tackle.

Comparing the security promises for MESO-based static and dynamic camouflaging (Table 3.5), it is clear that the number of wrongly inferred gates is higher for dynamic camouflaging. This increase also translates to a higher HD (about 5.14%). The OER, however, remains at 100% for both the schemes. Figure 3.14 shows the dependence of *HackTest*'s success rate as a function of HD for selected ITC-99 benchmarks. This plot reiterates that the degree of functional reconfiguration (measured as HD) has a strong impact on the accuracy of *HackTest*.

Table 3.4 Impact of increasing the number of possible functions implemented by the MESO-based primitive on *HackTest*'s accuracy for selected ITC-99 benchmarks. Test patterns are generated by *Tetramax ATPG* for fault coverage and test coverage of 99% and 100%, respectively. Results are averaged across 10 random trials of camouflaged gates

Benchmark	# Camo. gates	3 functions	4 functions	8 functions	16 functions
b14_C	350	20.37	15.13	14.69	11.49
b15_C	350	11.47	10.4	8.58	7.23
b20_C	350	17.03	14.03	11.11	8.51
b22_C	350	27.03	21.52	15.33	12.48
Average	350	18.98	15.27	12.43	9.93

Table 3.5 Impact of *HackTest* for MESO-based static (S. Camo) and dynamic camouflaging (D. Camo) schemes, on HD and OER for selected ITC-99 benchmarks. The number of wrongly inferred gates for dynamic camouflaging is higher on average when compared to static camouflaging, which translates to an improved HD. HD and OER are calculated by averaging across 10 random trials

	# Wrongly inferred gates		HD (%)		OER (%)	
Benchmark	S. Camo.	D. Camo.	S. Camo.	D. Camo.	S. Camo.	D. Camo.
b14_C	249	298	36.04	42.09	100	100
b15_C	296	320	32.07	34.23	100	100
b20_C	279	310	32.15	35.03	100	100
b22_C	212	296	20.09	29.57	100	100
Average	259	306	30.09	35.23	100	100

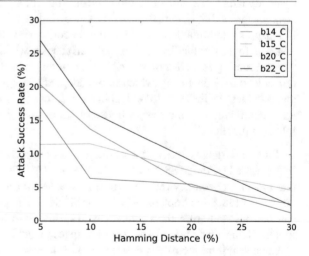

Fig. 3.14 Success rate of *HackTest* as a function of HD for selected ITC-99 benchmarks. It is evident that the degree of functional reconfiguration, also expressed by HD, can be leveraged by a designer to influence the overall success rate for *HackTest*

3.4.5 Dynamic Camouflaging Against Untrusted End-User

Malicious end-users of the IC pose a serious threat to safeguarding the proprietary design IP from RE. This risk is arguably even greater than those from untrusted foundries and test facilities, because those facilities can be vetted and audited.

However, once the IC is in the hands of the end-user, the opportunities for attacks are manifold.

Threat Model   Typically, the threat model considered for an untrusted end-user is as described in the literature [SRM15, MGT15, Li+16]:

- The attacker has access to advanced, specialized equipment for reverse engineering an IC, which includes setup to depackage an IC, delayer it, imaging of individual layers, and image processing tools.
- Further, he/she can readily distinguish between a camouflaged cell and a regular, standard cell. If hybrid spin-CMOS circuits are used, it is straightforward to identify the CMOS gates, whereas the complexity is increased manifold, if all the gates are implemented using MESO devices.
- The attacker is aware of the total number of camouflaged gates, and the number and type of functions implemented by each camouflaged cell.
- He/she procures multiple chip copies from the open market, uses one of them as an oracle (to observe the input–output mapping), and extracts the gate-level netlist of the chip by reverse engineering the others. This paves the way for algorithmic SAT-based attacks [SRM15, MGT15, Sha+17a].
- The attacker cannot invasively probe the output of a camouflaged cell. It is straightforward to note that once an adversary is allowed probing capabilities, i.e., to probe the output of a camouflaged cell, then the security guarantees offered by these schemes are substantially weakened.

Attack Setup   In 2015, Subramanyan et al. [SRM15] and Massad et al. [MGT15] independently demonstrated SAT-based attacks to circumvent security guarantees offered by LL and LC, respectively. Interested readers are referred to the respective papers for further details. This case study leverages the publicly available attack [SRM15] to perform the security evaluation for an untrusted end-user, as is the standard practice in the hardware security research community.

MESO-Based Polymorphic Circuits   The effectiveness of dynamic camouflaging against attacks arising from the perspective of malicious end-users can be understood through a conceptual example. Consider the circuit in Fig. 3.15. Here, $X4$ is the only camouflaged, polymorphic gate modeled with three key bits. Assuming a key distribution such that INV, BUF, AND, OR, NAND, NOR, XOR, and XNOR gates correspond to key bits {000, 001, ..., 111}, respectively, the dynamic key of the circuit cycles from 100 to 101, and then to 111, as per the outlined functional reconfiguration in Fig. 3.15.

The application of SAT-based attacks [SRM15, MGT15] for the simple scenario in Fig. 3.15 is explained next, through the attack iterations outlined in Fig. 3.16. Consider that the oracle (i.e., an actual working chip obtained from the market) implements f_1 during the first iteration of the SAT solver, where the input applied is 101. Note that the oracle is to be configured for test mode, to provide access to the circuit internals through scan chains, as required when modeling the whole circuit for the SAT-based attack. In principle, the oracle may behave differently in

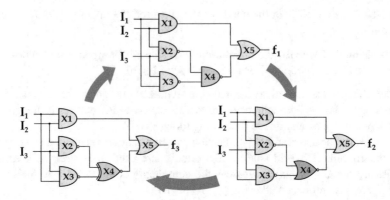

Fig. 3.15 Dynamic morphing of gate X4 in a representative circuit. Circuit implementing f_1 is the original template and f_2, f_3 are the morphed versions

Input $I_1I_2I_3$	Oracle output $f_1f_2f_3$	Current oracle	Output for different key combinations								Inference
			k_0	k_1	k_2	k_3	k_4	k_5	k_6	k_7	
000	001		0	1	1	1	0	0	0	1	
001	001		0	1	1	1	0	0	0	1	
010	001		0	1	1	1	0	0	0	1	
011	100	f_2	1	0	0	1	1	0	1	0	iter 2: k_0, k_3, k_4, k_6 pruned
100	001		0	1	1	1	0	0	0	1	
101	100	f_1	0	1	0	1	1	0	1	0	iter 1: k_0, k_2, k_5, k_7 pruned
110	111		1	1	1	1	1	1	1	1	
111	111		1	1	1	1	1	1	1	1	

Fig. 3.16 SAT-based attack [SRM15] on the polymorphic circuit of Fig. 3.15. For k_0 and k_1, the INV and BUF operations are performed on the output of $X2$

the test mode and in the operational (functional) mode. Naturally, the SAT solver is oblivious to the function being active internally in the oracle during *any* iteration. Also note that, once inputs are applied to the oracle, the SAT solver has to wait until the oracle provides the corresponding outputs. Now, that first SAT iteration prunes key combinations k_0, k_2, k_5, and k_7. While this is happening, assume that the gate $X4$ has morphed into NOR, and the oracle is now implementing function f_2. In the second SAT iteration, the input pattern 100, therefore, eliminates keys k_3, k_4, and k_6. Thereafter, the SAT solver concludes that the correct key bit and identity of gate $X4$ are 001 and BUF, respectively.

In essence, dynamic camouflaging can deceive and mislead the SAT solver to converge to an incorrect key, leading to an incorrect gate assignment. Consider for an exploratory study, an extension of the framework from [SRM15] to realize SAT-based attacks on polymorphic versions of ITC-99 benchmarks. Even for 100,000 randomized trials, the related attacks fail due to inconsistent I/O mappings, as these induce unsatisfiable (*UNSAT*) scenarios for the attack framework.

3.5   Closing Remarks

Functional polymorphism has been largely unexplored in the context of securing hardware. Dynamic camouflaging, a novel design-for-trust technique, built on the foundations of runtime polymorphism and post-fabrication reconfigurability exhibited by emerging devices, could be the key to securing the IC supply chain end-to-end. In this chapter, we delved into how emerging devices can transform the future of secure hardware, by enabling the obfuscation of circuits via dynamic morphing. In particular, we focused on the primary considerations for such emerging devices to be molded into dynamic camouflaging primitives. We explored the various emerging device candidates for implementing such schemes in future chips. Further, we investigated every aspect of designing and testing a dynamic camouflaging primitive against various adversaries in the IC supply chain.

References

[Ang+20] S. Angizi et al., Hybrid spin-CMOS polymorphic logic gate with application in in-memory computing. IEEE Trans. Magn. **56**(2), 1–15 (2020)

[BB16] S. Bauer, C.A. Bobisch, Nanoscale electron transport at the surface of a topological insulator. Nat. Commun. **7**(1), 1–6 (2016)

[Bi+16] Y. Bi et al., Emerging technology-based design of primitives for hardware security. J. Emerg. Tech. Comput. Syst. **13**(1), 3:1–3:19 (2016). ISSN: 1550-4832. https://doi. org/10.1145/2816818

[Der+11] H. Dery et al., Reconfigurable nanoelectronics using graphene based spintronic logic gates, in *Spintronics IV*, vol. 8100 (International Society for Optics and Photonics, 2011), 81000W

[DP71] M.I. Dyakonov, V.I. Perel, Current-induced spin orientation of electrons in semiconductors. Phys. Lett. A **35**(6), 459–460 (1971)

[KUH18] N.H.D. Khang, Y. Ueda, P.N. Hai, A conductive topological insulator with large spin Hall effect for ultralow power spin–orbit torque switching. Nat. Mater. **17**(9), 808–813 (2018)

[Li+16] M. Li et al., Provably secure camouflaging strategy for IC protection, in *Proc. Int. Conf. Comp.-Aided Des.* Austin, Texas (2016), pp. 28:1–28:8. ISBN: 978-1-4503-4466-1. https://doi.org/10.1145/2966986.2967065

[LO19] L. Li, A. Orailoglu, Piercing logic locking keys through redundancy identification, in *Proc. Des. Autom. Test Europe* (2019). https://doi.org/10.23919/DATE.2019.8714955

[Lot+04] T. Lottermoser et al., Magnetic phase control by an electric field. Nature **430**(6999), 541 (2004)

[Man+19] S. Manipatruni et al., Scalable energy-efficient magnetoelectric spin–orbit logic. Nature **565**(7737), 35–42 (2019)

[Mas+17] M. El Massad et al., Logic locking for secure outsourced chip fabrication: a new attack and provably secure defense mechanism. Comp. Research Rep. (2017). http://arxiv. org/abs/1703.10187

[MGT15] M. El Massad, S. Garg, M.V. Tripunitara, Integrated circuit (IC) decamouflaging: reverse engineering camouflaged ICs within minutes, in *Proc. Netw Dist. Sys. Sec. Symp.* (2015), pp. 1–14. https://doi.org/10.14722/ndss.2015.23218

[Moo+16] B. Moons et al., Energy-efficient convNets through approximate computing, in *2016 IEEE Winter Conference on Applications of Computer Vision (WACV)* (IEEE, Piscataway, 2016), pp. 1–8

[Pat+18d] S. Patnaik et al., Best of both worlds: Integration of split manufacturing and camouflaging into a security-driven CAD flow for 3D ICs, in *2018 IEEE/ACM International Conference on Computer-Aided Design (ICCAD)* (IEEE, Piscataway, 2018), pp. 1–8

[Pat+19a] S. Patnaik et al., A modern approach to IP protection and trojan prevention: split manufacturing for 3D ICs and obfuscation of vertical interconnects. IEEE Trans. Emerg. Top. Comput. (2019). https://doi.org/10.1109/TETC.2019.2933572

[Pat+20b] S. Patnaik et al., Obfuscating the interconnects: low-cost and resilient full-chip layout camouflaging. IEEE Trans. Comput. Aided Des. Integr. Circuits Syst. (2020). https://doi.org/10.1109/TCAD.2020.2981034

[Raj+13a] J. Rajendran et al., Security analysis of integrated circuit camouflaging, in *Proc. Comp. Comm. Sec.* Berlin, Germany (2013), pp. 709–720. ISBN: 978-1-4503-2477-9. https://doi.org/10.1145/2508859.2516656

[RFK19] S. Rakheja, M.E. Flatté, A.D. Kent, Voltage-controlled topological spin switch for ultralow-energy computing: performance modeling and benchmarking. Phys. Rev. Appl. **11**(5), 054009

[Sar+18] S.S. Sarwar et al., Energy-efficient neural computing with approximate multipliers. ACM J. Emerg. Technol. Comput. Syst. (JETC) **14**(2), 1–23 (2018)

[Sha+17a] K. Shamsi et al., AppSAT: approximately deobfuscating integrated circuits, in *Proc. Int. Symp. Hardw.-Orient. Sec. Trust* (2017), pp. 95–100. https://doi.org/10.1109/HST.2017.7951805

[SRM15] P. Subramanyan, S. Ray, S. Malik, Evaluating the security of logic encryption algorithms, in *Proc. Int. Symp. Hardw.-Orient. Sec. Trust* (2015), pp. 137–143. https://doi.org/10.1109/HST.2015.7140252

[SVR14] K. Shen, G. Vignale, R. Raimondi, Microscopic theory of the inverse Edelstein effect. Phys. Rev. Lett. **112**(9), 096601 (2014)

[SZK01] A. Stoica, R. Zebulum, D. Keymeulen, Polymorphic electronics, in *International Conference on Evolvable Systems* (Springer, Berlin, 2001), pp. 291–302

[Yas+16a] M. Yasin et al., Activation of logic encrypted chips: Pre-test or post-test?, in *Proc. Des. Autom. Test Europe* (2016), pp. 139–144

[Yog+15] K. Yogendra et al., Domain wall motion-based low power hybrid spin-CMOS 5-bit flash analog data converter, in *Sixteenth International Symposium on Quality Electronic Design* (IEEE, Piscataway, 2015), pp. 604–609

[YSR17] M. Yasin, O. Sinanoglu, J. Rajendran, Testing the trust-worthiness of IC testing: an oracle-less attack on IC camouflaging. Trans. Inf. Forens. Sec. **12**(11), 2668–2682 (2017). ISSN: 1556-6013. https://doi.org/10.1109/TIFS.2017.2710954

Chapter 4
Nonlinearity for Physically Unclonable Functions

4.1 Chapter Introduction

Authentication and other security schemes have, already for many decades, leveraged the randomized manifestations of selected physical, biological, or other
phenomena. For example, biometric identification and authentication are based
on the unique patterns of fingerprints, retinas, voices, or even walking pace, and
motion [TŠK07]. For electronic circuits, the corresponding notion of *physically
unclonable functions (PUFs)* has been established around 2002 [Pap+02, MV10,
Her+14, CZZ17]. A PUF should exhibit a device-specific and decorrelated, yet
reproducible, output behavior. That is, even under varying environmental conditions, the responses should be reproducible for the very same PUF instances
but should differ across different PUF instances and devices, even for the same
PUF design. The core working principle of PUFs is to leverage the process
variations inherent to microelectronics fabrication and operation and to boost these
variations purposefully. Prominent types of PUFs are ring oscillators, arbiters,
bistable rings, and memory-based PUFs [MV10, Her+14, Gan17, CZZ17]. Such
PUFs are relatively simple to implement and integrate in CMOS technology, even
for advanced processing nodes. However, for CMOS implementations of such
PUFs, the underlying variations and the related PUF properties (i.e., uniqueness,
unpredictability) can be limited and may be predicted. In fact, various attacks have
been demonstrated on CMOS PUFs, with machine learning emerging as the most
powerful approach [CZZ17, Rüh+13a, Liu+18, Gan17]. In contrast, PUFs built
from emerging technologies can be more promising, as they may leverage harder-to-
predict variations and randomness of emerging technologies that often exhibit some
inherently nonlinear behavior.

In this chapter, we first discuss concepts for PUFs using emerging technologies.
Then, we review selected prior art of PUFs using emerging technologies. Finally,
we present a case study on plasmonics-enhanced optical PUFs.

N. Rangarajan et al., *The Next Era in Hardware Security*,
https://doi.org/10.1007/978-3-030-85792-9_4

4.2   Concepts for Physically Unclonable Functions Using Emerging Technologies

The desired properties for PUFs are uniqueness, unclonability, unpredictability, reproducibility, and tamper-resilience. The following provides some background and context for emerging technologies:

1. **Uniqueness:** For the same PUF concept and design, different instances are expected to behave differently, i.e., to provide different responses, even for the same inputs. Uniqueness is commonly quantified by Hamming distances (HD) for outputs of multiple PUF instances; the ideal HD value for uniqueness is 50%. Uniqueness is limited by the particular PUF design as well as the device technology and inherent variations. For traditional CMOS technology, the underlying process variations are typically well controlled, which inherently limits CMOS-based PUFs to some degree. In contrast, by their heterogeneous and more varied nature, emerging technologies can offer various sources of process variations. For example, Wang et al. [WYM14] propose combining different PUF designs for 3D ICs, and Smith et al. [SS17] leverage metallic nanomaterials for optical tagging.

2. **Unpredictability:** PUFs should behave truly randomly, i.e., without any bias. This property has become more important over the years, given the rise of powerful machine learning frameworks that are well capable of learning and predicting various kinds of PUFs. Truly random behavior can be rooted in the underlying device itself; emerging technologies appear promising toward that end, as demonstrated in more detail in this chapter as well as in Chap. 5.

3. **Unclonability:** This is the key property for PUFs. It represents a somewhat abstract property by itself, but its practical implications are to be carefully studied for any particular PUF of interest. More specifically, unclonability bifurcates into (1) physical unclonability, as in impossibility to manufacture the very same PUF instance multiple times, and (2) mathematical unclonability, as in impossibility to fully model the behavior of a particular PUF instance. Truly unclonable PUFs need to fulfill both aspects, and there are few PUF approaches or technologies that may truly meet these aspects (e.g., those based on quantum physics). Nevertheless, the resilience of particular PUFs can be extraordinary in practice, which is then often considered sufficient for most security requirements. With the help of emerging technologies, the notion of unclonability may be improved over state-of-the-art CMOS-based PUFs. For example, memristors make good candidates for PUFs, given their underlying stochastic operation with nonlinear conductance variations and post-manufacturing tunability [Nil+18].

4. **Reproducibility:** This property can also be understood as reliability, i.e., whether one is able to observe the same response/behavior for the same PUF instance. Reproducibility is commonly quantified by HD for the same PUF instance; the ideal HD for reproducibility is 0%. Reproducibility can be significantly undermined by ambient conditions, like temperature and voltage

variations or electromagnetic interference, but also by time-dependent phenomena like aging of the underlying devices. These concerns are prominent already for CMOS-based PUFs and, depending on the maturity of the manufacturing processes, even more so for emerging technologies. Still, this statement cannot be generalized and each PUF design has to be studied for reproducibility.

5. **Tamper-resilience:** This property is typically understood as follows. Any tampering of a PUF instance is likely to affect the PUF itself; such tampering should in principal be detectable through alterations of the PUF behavior. Still, tampering would naturally impact the other properties as well, and it is important to understand which type and degree of tampering would undermine these properties without being readily detectable. For example, fault injection is a concern for most CMOS-based PUFs, where the attacker has relatively good control of the degree of faults and resulting bias being introduced for the PUFs, which can allow an attacker to subsequently better learn the PUFs' behavior [Taj17]. Some emerging technologies, like phase change memories, can offer inherent tamper-resilience as well [Rah+17].

Most if not all PUF implementations, irrespective of whether they are based on CMOS or emerging technologies, have some limitations for these properties. Still, emerging technologies appear more promising than CMOS technology to meet these properties, albeit with inherent trade-offs becoming more relevant. For example, process variations tend to be more pronounced for most emerging technologies— such stronger variations serve well for uniqueness/entropy of PUFs, but may hinder reproducibility at the same time. Overall, any PUF design, irrespective of their use of CMOS or some emerging technology, requires detailed studies, initially for the conceptual and simulation level as well as subsequently for prototypes and measurements.

4.3 Review of Selected Emerging Technologies and Prior Art

4.3.1 Memristive Devices

The potential for using memristors toward hardware security schemes has been recognized already some years ago, e.g., for PUFs leveraging the process variations and the stochastic operation of memristors in 2013 [Ros+13].

More recently, another memristive device-based PUF has been proposed [Nil+18], which leverages the nonlinear I-V characteristics of memristors ("pinched hysteresis") and applies analogue tuning of the memristor conductance. This is done to increase the performance and practicality of such PUFs and to reduce the complexity of the peripheral circuitry. The authors of [Nil+18] provided an experimental demonstration and measurement results for their PUF concept. More specifically, their demonstrated circuit uses two vertically integrated 10×10 metal oxide memristive crossbar circuits (Fig. 4.1). As with memristor devices in

Fig. 4.1 Concept of a
memristive crossbar PUF, as
proposed in [Nil+18].
Memristors are embedded at
the junctions of rows and
columns

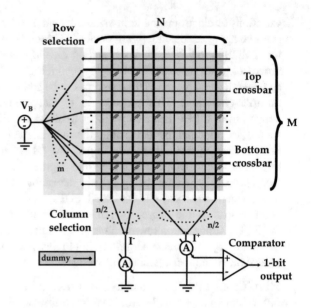

general, the manufacturing is compatible with back-end-of-line processing along
with regular CMOS manufacturing. The measurement results indicate a uniqueness
near ideal 50%, as well as high reproducibility or low bit-error rates ($1.5 \pm 1\%$).
The authors furthermore conduct machine learning-based attacks, indicating strong
resilience against such attacks as well (owing to the underlying nonlinearities).

4.3.2   Carbon Nanotube Devices

In [LHH18], the authors propose PUFs that leverage the manufacturing variability of
CNTs, along with the notion of Lorenz chaotic systems. The latter serves to enhance
the decorrelation of inputs and outputs and, thus, renders these PUFs more resilient
against machine learning attacks.

More specifically, the authors propose a crossbar structure with CNTs at its
heart and their imperfections serving as a source of randomness. The crossbar
structure is augmented with digital to analog converters (DACs) for inputs and
vice versa for outputs, as well as current measurement and comparator circuitry
(Fig. 4.2). Accordingly, the input/output behavior is mapped from the digital domain
to the physical, as threshold-driven currents through imperfect CNTs, and back for
evaluation. While this basic crossbar structure is difficult to clone by manufacturing,
it is relatively easy to clone by modeling. Accordingly, an important subsequent
stage is a Lorenz chaotic system module, which introduces the necessary resilience
against machine learning attacks and others. A Lorenz chaotic system exaggerates
the differences for output response across similar inputs. The strength/degree of

Fig. 4.2 Concept of a CNT crossbar PUF, along with signal processing based on Lorenz chaotic system, as proposed in [LHH18]. CNTs are embedded at the junctions of rows and columns. DACs and ADCs, along with comparator circuitry, are used for challenging and reading the PUF

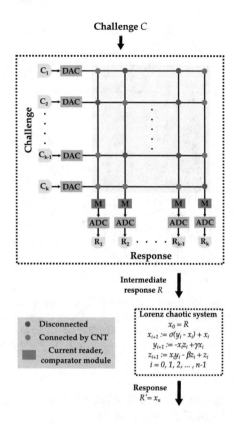

this chaotic behavior is dictated by the system parameters; for better resilience, the authors derive these from the intermediate response coming from the CNT crossbar structure itself. In their experimental evaluation, the authors demonstrate that such a compound scheme (which could be well leveraged for other devices aside from CNT crossbar structures) limits various machine learning models to around 55% bit prediction, which is only marginally better than random guessing.

Aside from the particular PUF proposal above, we note that in [Rah+17], the authors review the use of CNTs for PUFs, TRNGs, and also propose the technology to be used for novel sensors for detecting microprobing or other invasive attacks.

4.3.3 3D Integration

The integration of multiple chips into 3D/2.5D stacks—discussed in more detail in Chap. 6—also seems beneficial for advancing the notion of PUFs. This is because the individual chips/active layers within such stacks are subject to independent process variations. One can build up PUFs that combine these multiple sources of

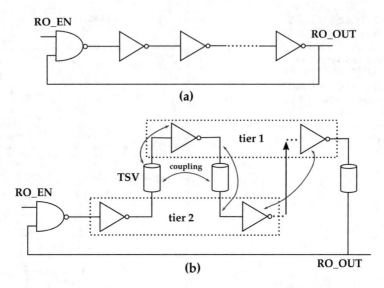

Fig. 4.3 Ring oscillator PUFs. (**a**) Regular structure, implemented in 2D ICs. (**b**) Advanced structure, implemented in 3D ICs, as proposed in [Wan+15a]. The oscillator comprises active devices across multiple chips (dashed boxes) as well as TSVs (cylinders), which all contribute their individual variations. Furthermore, there are coupling phenomena across the oscillator components (arrows)

entropy at the 3D/2.5D system level, along with additional coupling effects arising for 3D/2.5D stacks as well.

For example, in [WYM14, Wan+15a], two such schemes have been proposed. These schemes are based on ring oscillators [Wan+15a] (Fig. 4.3) and clock-skew arbiter structures [WYM14], respectively. While generic in principle, the scheme in [WYM14] suggests 3D integration as a particularly promising implementation option, whereas [Wan+15a] explicitly leverages, aside from regular process variations, the process variations of through-silicon vias (TSVs), i.e., large metal plugs running through chips in their entirety, to interconnect these chips across the 3D stack. Based on technology simulation, the authors of [Wan+15a] find that their scheme improves both uniqueness as well as reproducibility, i.e., when compared to various other PUF architectures implemented in regular 2D ICs.

Although promising in principle, these studies did not consider state-of-the-art machine learning attacks. While one may expect some increase of resilience from the compounding action of multiple entropy sources, the key question is whether such entropy is only linearly superimposed or intertwined in a more complex manner. Given the fact that there are various physical phenomena in 3D/2.5D stacks that are intertwined across chips/layers (e.g., various coupling effects for TSVs, active devices, substrate, wires across the stack; nonlinear heat conduction paths due to heterogeneous material composition; thermo-mechanical stress induced by

TSVs that impacts carrier mobility; et cetera), there is some promise, but detailed studies are required in any case.

4.3.4   Optical Devices

There are various studies as well as prototypes for PUFs based on optical phenomena [Pap+02, Rüh+13b, TŠ07, MV10, Gru+17b]. The most commonly pursued approach is to manufacture an "optical token." In addition to structural variations inherently present in the optical media of the tokens, these may further contain randomly included materials, e.g., microscopic particles. The fundamental underlying phenomena of an optical PUF are scattering, reflection, coupling, and absorption of light within the optical token. Depending on the materials used for the token and the inclusions, as well as the design and structure of the token itself, these phenomena can be highly nonlinear and chaotic by nature [Gru+17b, KZ12].

In 2002—for the very first PUF proposal in the literature—Pappu et al. [Pap+02] devised an optical token from transparent epoxy with randomly inserted, micrometer-sized glass spheres. That token was illuminated by an external laser, whereupon the resulting speckle pattern was visually recorded, filtered, and digitized (Fig. 4.4). In 2013, Rührmair et al. [Rüh+13b] first replicated and confirmed the findings by Pappu et al. and then prototyped an integrated optical PUF based on the same working principle. Tuyls et al. [TŠ07] discussed integrated optical PUFs in 2017, albeit only in theory, without any experimental evaluation. Also in 2017, Grubel et al. [Gru+17b] demonstrated a resonator-based PUF with pseudo-randomized structures but with inherently nonlinear behavior, due to the use of silicon as optical medium.

Prior art on optical PUFs has some practical limitations, e.g., the use of linear media, external and exposed optical tokens, the need for complex and sensitive setups, or the need to customize manufacturing steps for different PUF tokens. More specifically, concerning the early works on external optical PUFs [Pap+02, Rüh+13b], a major shortcoming is the use of linear materials that can be modeled [Rüh+13b]. Furthermore, their respective setups are relatively complex. Thus, these PUFs are not only sensitive to environmental parameters like variations of

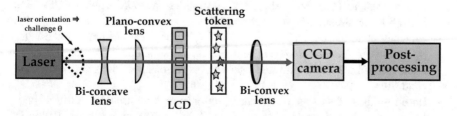

Fig. 4.4 Concept for an optical PUF, as proposed in [Pap+02] and verified in [Rüh+13b]

temperature and supply voltages, which is the case for any type of PUF, but also to mechanical vibrations, laser alignment, etc. Besides, exposing the token can result in wear and tear; an external PUF may become irreproducible after some time. Even more concerning, an external PUF can arguably never be completely trusted—an attacker can take hold of the token and, subsequently, (a) reuse it for authenticating of counterfeit chips, or (b) explore its challenge-response behavior for modeling attacks. The more recent waveguide PUF [Gru+17b], while certainly an advancement, still has limitations. For one, it requires sophisticated external optoelectronic components, e.g., pulse pattern generators and programmable spectral filters. For another, this PUF relies on pseudo-randomized structures within the resonator; this requires customizing the manufacturing steps for different PUFs that may be impractical.

4.4   Case Study: Plasmonics-Enhanced Optical PUFs

Aside from optical PUFs, it is noteworthy that particles of different types and sizes have been used for some time for "tagging" and optical authentication of goods [RDK12, SS17]. In fact, the first well-known approach for secure authentication of goods was particle-based tagging of nuclear weapons during the cold war [RDK12]. Different materials can be leveraged for particle-based tagging, e.g., quantum dots, fluorescent particles, or metallic nanoparticles (NPs). The latter are particularly interesting, as they effectively boost the nonlinear light–matter interaction [KZ12, Par+16], which is extremely valuable when seeking to render a PUF resilient against modeling attacks [Gan17]. Indeed, the concept of *plasmonic NPs* has gained significant traction for security schemes [SS17, Zhe+16, SPS16, Par+16, Cui+14]. As with most optical PUFs, however, current schemes require external components; integrated schemes have not been proposed yet.

In this case study, we describe the work by Knechtel et al. [Kne+19] in detail. The key idea is to consider the nonlinear phenomena of silicon photonics and plasmonics at once. The scope of this case study can be summarized as follows:

- Knechtel et al. propose to entangle two highly nonlinear physical phenomena by construction: light propagation in a silicon disk resonator, and surface plasmons arising from NPs arranged randomly on top of the resonator. The authors name this concept as *peo-PUFs*, short for plasmonics-enhanced optical PUFs. See Fig. 4.5 for an illustration of the concept.
- Knechtel et al. discuss and study the underlying physical phenomena, the latter by means of *Lumerical FDTD* and *COMSOL* simulations. Using the same tools, the authors obtain different data sets for various peo-PUFs under different conditions.
- Based on these data sets, randomness, uniqueness, and reproducibility of peo-PUFs are evaluated, confirming their applicability for secure key generation in

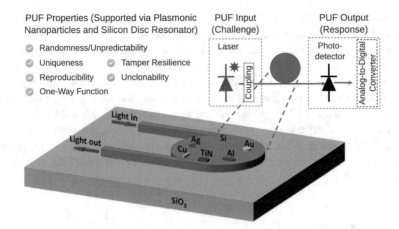

Fig. 4.5 Concept for peo-PUFs. At its heart is a device-integrated, micrometer-sized silicon disk resonator with plasmonic NPs of different shapes, sizes, and metals, randomly arranged on top of the resonator

principle. Knechtel et al. also propose a simple authentication scheme, based on the secure keys and some related helper data.
- The concept can be directly integrated into any silicon device; thus it can provide better resilience than prior external PUFs. The manufacturing of silicon disc resonators is a well-known process, and plasmonic NPs can be easily deposited on top of a resonator in a random fashion, e.g., by sputtering. Knechtel et al. argue that these two aspects are essential for successful, future adoption of peo-PUFs.

While focusing on a particular concept and its evaluation, the idea of presenting such a case study here also provides guidance by example as to which steps a device expert would want to take to propose their own emerging device-based PUF.

4.4.1 Exploration of Device Physics

The first step in constructing an emerging device-based PUF primitive is to study the physical phenomena underlying the device of interest and to see how these map to the outlined concepts of PUFs (Sect. 4.2).

4.4.1.1 Light Propagation in Silicon Resonators

It is well-known that silicon is transparent to infrared light and has a very high refractive index. The accordingly strong photonic confinement allows for the design of micrometer-scale, yet efficient, silicon-made optical waveguides [LF18]. Among others, prominent applications are optoelectronic filters based on *silicon disc resonators (SDRs)*, i.e., circular cavities that enable strong light interaction. That is, light entering an SDR builds up in intensity over multiple round-trips within the cavity, where only particular wavelengths, depending on the SDR design, will be in resonance.

The light propagation within SDRs in particular and silicon devices in general are subject to various nonlinear effects. These include the *Kerr effect, Raman scattering, self-modulation, two-photon absorption*, et cetera [LF18, KZ12, Gru+17b, Wan+13]. The nonlinearity in silicon is fundamental for light-to-light interaction, which itself is essential for techniques such as wavelength conversion. Depending on the SDR design, other chaotic effects like *dynamic billiards* can also play a role [Gru+17b].

4.4.1.2   Plasmonics

Because of the outstanding capability for sub-wavelength confinement of electro-magnetic energy, plasmonics has become a driving force for progress in the area of nanophotonics [Gos+19, GR19, Kum+13]. Metallic nanostructures are at the heart of plasmonics—the phenomenon of plasmonics originates from strong coupling of photon energy with free electrons in a metal. This strong coupling supports a wave of charge-density fluctuations along the surface of the metal, thereby creating a sub-wavelength oscillating mode called a *surface plasmon* [Boz09, Che+11, KZ12]. More specifically, for one, there is the electromagnetic energy transport for the propagation of *surface plasmon polaritons (SPPs)* sustained at the planar metal/dielectric interface; for another, there is the *localized surface plasmon resonance (LSPR)*, a non-propagating excitation of the metal's conduction electrons coupled to the electromagnetic field. Note that the LSPR in particular is under intensive research for many years now [WS06, Sch+10, Par+16, KZ12, Gos+17, SKD12]; that interest is also because a large variety of metallic nanostructures are commercially available, which all give rise to unique properties and applications.

The LSPR field enhancement is dictated by various factors. First and foremost, the enhancement depends on the metal properties [Gos+17]. Moreover, the field enhancement also depends on the NP structure and size [WS06, Sch+10, Hua+07, Nog07], the coupling arrangement or direction [Gos+17], and the materials surrounding the NP [SPS16]. See Fig. 4.6 for simulations with differently sized and arranged NPs; the simulation setup is described in Sect. 4.4.3.1.

Besides LSPR field enhancement, when light interacts with a NP, the light can be absorbed and/or scattered. Those processes arise in resonant conditions, i.e., where the absorption efficiency is the highest. The *extinction efficiency*, that is the sum of absorption and scattering efficiencies, can be engineered via the size, shape, and dielectric environment of the metallic NPs (Fig. 4.7; see below for details). For NPs smaller than the wavelength of excitation, note that the efficiency of absorption dominates over scattering. For NPs that have one or more dimensions approaching the excitation wavelength, the optical phase can vary across the structure. Thus, the retardation effect should be accounted for. Such nanostructures can also be

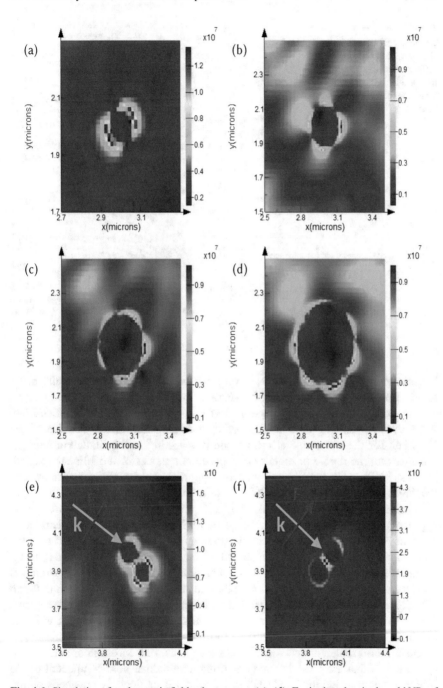

Fig. 4.6 Simulations for plasmonic field enhancement. (**a**)–(**d**): Excitations by single gold NPs of different sizes. (**e**), (**f**): Excitations by two gold NPs (same size) that are (**e**) not coupled versus (**f**) coupled. The latter shows a strong non-resonant gap mode. Vectors E are for the electric field and k for the light propagation, respectively

Fig. 4.7 Extinction
efficiencies for (**a**) gold and
(**b**) titanium nitride NPs of
different sizes, but uniform
height/thickness (30 and
20 nm, respectively). NPs are
labeled as "disk" in the plots

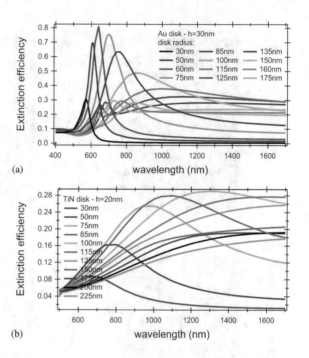

(a)

(b)

considered as SPP waveguides that propagate back and forth between the metal terminations, thereby creating a *Fabry–Perot resonator* for SPPs.

The extinction efficiencies illustrated in Fig. 4.7 were obtained via *COMSOL FEM* simulations. The electric field of light is polarized along the side of the NPs, i.e., for the transverse electric mode propagating in the SDR. For such an arrangement, the dipole or higher-order plasmonic modes of the NPs are excited; for small NPs, only the dipole mode is supported. As the NP radius increases, the higher-order modes appear, which shift the resonant conditions to longer wavelengths (see curve peaks). At the same time, higher-order modes can also appear at shorter wavelengths (e.g., Fig. 4.7a, for NP radius 135 nm). The extinction efficiency and resonance conditions depend not only on the NP size but also the material properties. The shape of the extinction efficiency curve depends mostly on the resonance conditions of an NP. For gold NPs (Fig. 4.7a), the extinction efficiency is narrow, with the maximum located at the wavelength of 640 nm, for an NP with a radius of 60 nm. In contrast, the extinction efficiency for titanium nitride NPs (Fig. 4.7b) is much broader and shifted toward longer wavelengths, with the maximum located at the wavelength of 1300 nm, for an NP with a radius of 100 nm.

Other non-resonant effects are to be considered as well. Most important, for two or more NPs that are close to each other, the longitudinal field component of the SPP can "jump inside the gap" such that the field is particularly enhanced within the gap (Fig. 4.6f). The enhancement magnitude depends on the size of the NPs and the spacing between them: the smaller the gap, the higher the enhancement. In case the field component is parallel to or outside of the gap, each of the NPs excites its "own" mode, which are not coupled (Fig. 4.6e).

4.4.2 Concept

The next step in constructing an emerging device-based PUF primitive is to devise a concept, including implementation aspects and also related security schemes. As for peo-PUFs, recall that the basic concept is illustrated in Fig. 4.5. Next, the implementation, working principle, as well as key generation, and authentication for peo-PUFs are outlined.

4.4.2.1 Implementation

It is important to note that, as of now, this case study is carried out at the level of modeling, physical simulations, and analytical security evaluation. For future work, the manufacturing and characterization of peo-PUFs as well as the verification of security promises in manufactured peo-PUFs would be of interest. In this context, note the following:

- The manufacturing of silicon waveguides/resonators as well as the deposition of NPs is a well-established processes. Unlike most prior art for optical PUFs, peo-PUFs can, therefore, take full advantage of commercial manufacturing facilities. Moreover, the token can be integrated on the electronic device—this is essential for proper implementation of security schemes.
- One can argue that advances for optoelectronics may enable monolithic integration sooner than later. For example, IBM has recently demonstrated an ultra-fast photonic intra-chip communication link [Xio+16]. In general, the monolithic integration of optical modulators and photodetectors is considered mature, with research focus shifting toward the integration of the light source/lasers [Sei+18, Gua+18, Orc+12, Sor06]. Besides, hybrid integration of plasmonics and silicon photonics has been demonstrated as well, e.g., by Chen et al. [Che+18].
- As indicated, while peo-PUF tokens can support broad ranges of optical inputs in principle, generating these inputs would require, e.g., spectral encoding, frequency sweeping, or power modulation. Therefore, the challenge here would be to manage the complexity of the related optoelectronics, which may render a monolithic integration of peo-PUFs ultimately more difficult. Hence, in the remainder, a simple setup with a fixed laser input pulse is assumed for the peo-PUFs.

4.4.2.2 Working Principle

Essentially, once an optical pulse enters the SDR token, the nonlinear phenomena
of silicon photonics and plasmonics shall result in a highly complex and chaotic, yet
reproducible, optical output. This output is then to be post-processed following the
security scheme of choice; see Sect. 4.4.2.3 for the latter for this case study.

Considering all the effects outlined in Sect. 4.4.1, it is intuitive that metallic
NPs on top of an SDR will significantly impact the photonic mode propagation
of the SDR. This, in turn, will help to induce highly nonlinear behavior for the peo-
PUFs. As outlined, the disturbances arise primarily through absorption, scattering,
and LSPR field enhancement, but with all effects acting at once. Given that the
SDR will be made from silicon, a nonlinear material, the field enhancement can
also locally impact the SDR's refractive index, thereby further disturbing the mode
propagation. As with plasmonic field enhancement, this interference depends on the
size of the NPs, their coupling direction, light polarization, material properties, et
cetera [SPS16].

Figure 4.8 illustrates the electric field and the photonic propagation mode for an
SDR acting as peo-PUF token. Parameters for the pulse are the same as in Figs. 4.6
and 4.9; see also Sect. 4.4.3.1. The SDR is 3.3 μm in radius and 180 nm thick. Only
for Fig. 4.8b, a gold NP (20 nm thick, 60 nm radius) is placed on top of the SDR (at
$x = -2$ μm and $y = -2$ μm). This arrangement allows to excite a dipole plasmonic
mode, illustrated as inset.

Figure 4.9 plots the pulse transmission for different arrangements of single NPs
on top of an SDR. For simplicity, and also to showcase the inherent potential for
highly nonlinear behavior of peo-PUFs, these simulations consider (a) only one
single gold NP versus no NP, and (b) no manufacturing variabilities, i.e., the SDR
and the NP are assumed to be perfect disks without any roughness, etc. Still, it

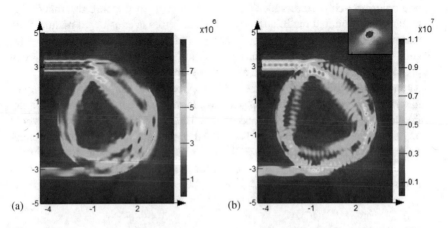

Fig. 4.8 Electric fields and photonic propagation for an SDR without NPs (**a**) versus with one
gold NP (**b**). The inset in (**b**) shows the propagation mode for the NP. Dimensions in μm

Fig. 4.9 Transmission plots for peo-PUF tokens with SDR radius 5 μm and height/thickness 180 nm. SDR tokens are independently simulated for three variations of one gold NP, regarding the NP sizes/radius. Without loss of generality, the considered NP radii are 60, 120, and 240 nm. For fair comparison, the NPs are always placed at the same location on the SDR

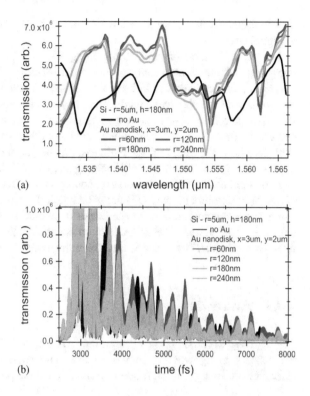

is evident that already one perfect NP has a significant impact on the propagation modes. In reality, manufacturing variabilities and more NPs can further enhance the randomness of peo-PUFs, thereby increasing their unpredictability.

Detrimental effects such as device aging, laser noise, or temperature variations may play a role in practice, but this can only be properly investigated once peo-PUFs are manufactured. In any case, error correction schemes to mitigate environmental impact on PUFs have been discussed in detail, e.g., in [Mae+15, Her+14, Rüh+13b].

4.4.2.3 Key Generation and Authentication

After applying a fixed input pulse, the output is processed as follows to obtain a key. First, wave shaping/spectral filtering is applied to extract different partitions (or features) across the whole wavelength spectrum of the optical response. Second, analog to digital conversion is applied on each feature to obtain the underlying bits. To enable a stable key, the noisy, least significant bits are rejected across all features. Finally, for each feature, their stable bits are bundled, and these bundles are grouped. Without loss of generality, the above steps are taken such that the key is 128 bits long. In general, the number of key bits dictates the number of bits to extract per feature and the number of features to be considered. This, in turn, dictates the requirements for accuracy and resolution of the optoelectronics and the analog to digital circuitry.

For authentication of chips using integrated peo-PUFs, the following simple two-phase scheme is proposed, following the literature [MV10]. During the initial *enrollment phase*, which is to be conducted in a trusted environment, the unique PUF responses are observed and different keys are derived. Besides the actual keys, some key parameters—also known as *helper data*—are to be recorded as well. For peo-PUFs, that is: (a) the selection of frequency features, and (b) the number of considered bits for each feature. Note that this helper data is essential to verify the keys. Furthermore, the *fractional Hamming distance (FHD)* is recorded for the same key obtained under different operating conditions. As for the actual *authentication phase*, after enrollment, the PUFs are to be queried again, using the same fixed input and a selected configuration for the helper data, and the resulting key is compared with the recorded one. The authentication is considered successful in case the key falls within the FHD expected for the operating conditions and helper data.

4.4.3   Evaluation

The final step in constructing an emerging device-based PUF primitive is to evaluate the concept along with its implementation and security schemes. This section elaborates on such simulation-based evaluation for the peo-PUFs.

4.4.3.1   Experimental Setup

Simulations for the frequency and time domain are carried out using the *Lumerical FDTD* software. All simulation data is exported for the subsequent security evaluation. If not specified otherwise, and without loss of generality, Knechtel et al. assume a 5 μm radius and a 180 nm height for the SDR, a height of 30 nm for the NPs, an optical input pulse of 100 fs duration, an ambient temperature of 300 K, and a simulation time-frame of 8 ps. Note that each simulation took multiple days; more NPs, longer simulation time-frames, etc., will significantly increase runtime. For absorption and scattering phenomena (Fig. 4.7), *COMSOL FEM* simulations were performed.

For the key generation and authentication framework, Knechtel et al. leverage *Matlab* and custom scripts. The security evaluation is based on *entropy*, *NIST randomness tests* [Bas+10], and FHD [MV10]. Without loss of generality, the authors leverage the frequency-domain data obtained from *Lumerical FDTD* simulations for the security evaluation. Knechtel et al. apply the post-processing steps outlined in Sect. 4.4.2.3. They sample 1000 different keys for each peo-PUF configuration under consideration, by varying the selection of frequency features and the number of bits per feature. Note that the authors release their post-processing and security evaluation framework to the community via [BPK19].

Table 4.1 Considered peo-PUF configurations

Label	Short description
(1) Si5umAu60nm(a)	SDR radius 5 μm, 1 gold NP radius 60 nm
(2) Si5umAu60nm(b)	As Si5umAu60nm(a), but different locations for NP
(3) Si5umAu60nm(c)	As Si5umAu60nm(a), but different locations for NP
(4) Si5umAu60nm(5)	As Si5umAu60nm(a), but 5 gold NP
(5) Si6umAu60nm	SDR radius 6 μm, 1 gold NP radius 60 nm
(6) Si5umAu120nm	SDR radius 5 μm, 1 gold NP radius 120 nm
(7) Si6umAu120nm	SDR radius 6 μm, 1 gold NP radius 120 nm
(8) Si5umAu240nm	SDR radius 5 μm, 1 gold NP radius 240 nm
(9) Si6umAu240nm	SDR radius 6 μm, 1 gold NP radius 240 nm
(10) Si5um	SDR radius 5 μm, no NP
(11) Pulse100fs	As Si5umAu60nm(a), but different locations for NP
(12) Pulse50fs	As Pulse100fs, but input pulse width 50 fs
(13) Pulse200fs	As Pulse100fs, but input pulse width 200 fs
(14) Temp300K	As Si5umAu60nm(a), but different locations for NP
(15) Temp350K	As Temp300K, but ambient temperature 350 K
(16) Si5umTiN60nm	As Si5umAu60nm(a), but 1 titanium nitride NP
(17) Si5umTiN60nm(5)	As Si5umAu60nm(a), but 5 titanium nitride NP
(18) Si5umTiN(5)	As Si5umAu60nm(a), but 5 titanium nitride NP with radius ranging from 60 to 240 nm

Knechtel et al. consider various peo-PUF setups, mainly for different arrangements of NPs, but also for different SDR tokens and operating conditions (temperature and pulse width). The configurations are labeled and summarized in Table 4.1.

4.4.3.2 Randomness

For peo-PUFs acting as "weak" PUFs,[1] it is essential to quantify the underlying randomness, which also reflects on their *unpredictability* [MV10]. As indicated, the entropy is evaluated and NIST tests are conducted in this case study toward that end.

---

[1] As indicated, "weak PUFs" are not necessarily inferior to "strong" PUFs [Rüh+13a]. On the contrary, powerful machine learning attacks such as [CZZ17, Rüh+13a, Liu+18, Gan17, Ata+18] do *not* apply for weak PUFs, only for strong PUFs. In any case, peo-PUFs may also be implemented as strong PUFs; the SDR tokens by themselves can readily support a large range of optical inputs. However, this would necessitate more complex optoelectronics for the PUF devices. In anticipation of the just outlined security concerns for strong PUFs, it should be noted that Atakhodjaev et al. [Ata+18] have shown that machine learning attacks on strong SDR-based PUF can be already challenging, due to the inherent nonlinearity of silicon SDRs. As motivated in this case study, once the additional phenomena of plasmonics become intertwined, one can reasonably expect even better resilience against such attacks.

Table 4.2 Entropy and passing of NIST tests

	Min S	Mean S	F (%)	FB (%)	R (%)	AE (%)
Si5umAu60nm(a)	0.933	0.9911	83.8	98.1	99.8	98.9
Si5umAu60nm(b)	0.9625	0.9972	98.6	100	83.9	96.5
Si5umAu60nm(c)	0.928	0.9871	72.3	99.1	99.9	98.1
Si5umAu60nm(5)	0.9575	0.9968	98.2	99.4	99.8	99.3
Si6umAu60nm	0.9727	0.997	98.5	99.9	98.3	98.8
Si5umAu120nm	0.9579	0.9969	98.4	100	93.1	98
Si6umAu120nm	0.9597	0.9979	98.6	100	98.7	98.7
Si5umAu240nm	0.9682	0.9961	96.3	99.2	99.1	98.4
Si6umAu240nm	0.9523	0.9963	97	99.6	100	99.8
Si5um	0.9259	0.9926	88.1	99.4	99.9	98.5
Pulse100fs	0.964	0.9964	97.3	100	90.9	97.4
Pulse50fs	0.9681	0.9978	98.6	100	83.8	96.5
Pulse200fs	0.9696	0.9965	98.8	99.7	100	99.3
Temp300K	0.9632	0.996	97	99.8	97.7	97.5
Temp350K	0.9258	0.9905	81.7	99.7	100	98.8
Si5umTiN60nm	0.9183	0.9895	80.6	98.7	99.7	97.9
Si5umTiN60nm(5)	0.9398	0.9902	83.6	99.2	99.4	98.2
Si5umTiN(5)	0.9423	0.9911	89.9	98.8	99.8	98.5

Knechtel et al. report the entropy S for the different peo-PUF configurations in Table 4.2. All mean entropies are beyond 0.987, which hints on strongly random distributions of zeroes and ones across all keys and for different peo-PUFs. For the minimal entropies, however, the authors note that *Si5um* and *Si5umTiN60nm*, i.e., the peo-PUF without any NP and the peo-PUF with one single titanium nitride NP, respectively, appear the weakest. Hence, the addition of plasmonic NPs can indeed help to increase randomness, but the degree of randomness depends on the NP count, materials, etc.

Knechtel et al. furthermore conduct the following, commonly considered NIST tests [Bas+10]:

1. Frequency (F): testing for the proportion of zeroes versus ones across the entire key
2. Frequency within a block (FB): testing for the proportion of zeroes versus ones within m-bit blocks, for $m = 20$
3. Runs (R): testing for the number of uninterrupted sequences (identical bits) across the entire key
4. Approximate entropy (AE): testing for the frequency of all possible m-bit patterns, with respect to an enumeration of all possible overlapping blocks of consecutive lengths (m and $m + 1$), for $m = 3$

The NIST tests are all based on *p-values*, i.e., they can quantify the confidence for passing (or failing) particular test. More specifically, the p-values represent the probability that a perfect random number generator would have produced a sequence less random than the sequence that was tested. In other words, if the p-value for a particular test is equal to 1, then the key appears to be perfectly random concerning the kind of randomness assessed by that test.

In Table 4.2, the percentage of keys passing the NIST tests is reported, all for a confidence interval of 99%. Knechtel et al. observe that the configurations *Si5umAu60nm*, *Si5um*, *Si5umAu60nm(c)*, and *Si5umTiN60nm* are inferior to others. Hence, configurations with none or only one NP are limited; already the positioning of single NPs may lead to these results. That is corroborated by the fact that the configuration *Si5umAu60nm(b)* is superior, although it also holds only one gold NP of same size and shape as the other configurations. In reality, peo-PUFs could and should comprise many more NPs, all randomly arranged. Knechtel et al. believe that the arrangement of an individual NP would play no significant, possibly deteriorating role there anymore for the overall randomness.

4.4.3.3 Uniqueness and Reproducibility

Two further key properties for any PUF are *uniqueness* and *reproducibility* [MV10]. Both uniqueness and reproducibility are to be measured on pairs of PUF outputs resulting from the same challenge. Uniqueness describes the difference of outputs across two PUF instances, whereas reproducibility describes the similarity of outputs for the same PUF instance, but under different operating conditions. Therefore, reproducibility can also be thought of as reliability. The FHD, short for fractional Hamming distance, is used to quantify both properties.

Regarding FHD for uniqueness, also known as *inter-FHD*, the ideal value is 50%; regarding FHD for reproducibility, also known as *intra-FHD*, the ideal value is 0%. Since the inter- and intra-FHD can vary depending on the applied challenge/helper data, Knechtel et al. report Gaussian FHD distributions along with their histograms, mean values μ, and standard deviations σ, as suggested in [MV10]. It is important to note that deviations from ideal inter-/intra-FHD values are tolerable as long as their distributions remain reasonably separated.

Next, peo-PUFs are investigated for two critical operation parameters, namely input pulse width and ambient temperature.

Knechtel et al. contrast the FHD distribution for different pulse widths in Fig. 4.10. That is, the authors assume that the input may exhibit some noise which peo-PUFs should be able to tolerate, at least to some degree. More specifically, the authors consider the scenario *Pulse100fs* versus *Pulse50fs* as tolerable fluctuations for one and the same input challenge, i.e., concerning reproducibility and intra-FHD. Another scenario, *Pulse200fs* versus *Pulse50fs*, is considered as comparing two different peo-PUFs with different laser setups, i.e., concerning uniqueness and inter-FHD. Now, from Fig. 4.10, one can note that the FHD distributions for these two scenarios are clearly distinct. Hence, the peo-PUFs are (a) reproducible for small input variations, around 50 fs, and (b) unique for different laser setups.

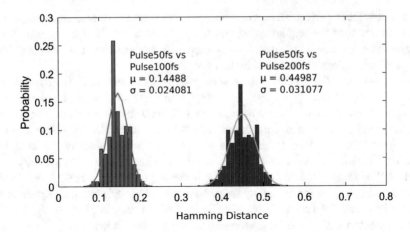

Fig. 4.10 Intra- and inter-FHD for different input pulses

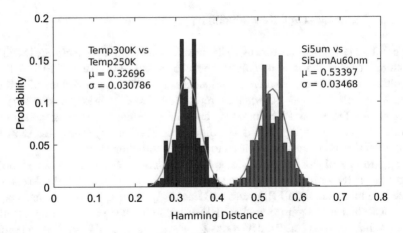

Fig. 4.11 Intra-FHD for different ambient temperatures but for the same peo-PUF, versus inter-FHD for different peo-PUFs. For the latter, note that the underlying spectra are provided in Fig. 4.9

The ambient temperature impacts the reproducibility of most, if not all, types of PUFs [MV10, CZZ17, Her+14]. In Fig. 4.11, the intra-FHD for the same peo-PUF (with one gold NP) is contrasted at 300 K versus 350 K ambient temperature, after applying correlation-based shifting of the wavelength spectra. While one can observe more noise than in the case for the reproducibility under input pulse fluctuations, the intra-FHD distribution still remains separated from another inter-FHD distribution (*Si5umAu60nm* versus *Si5um*) that was obtained from different peo-PUFs. Hence, peo-PUFs can tolerate some temperature fluctuations, although further compensation measures may be required in practice, where other noises such as voltage glitches may play some role as well.

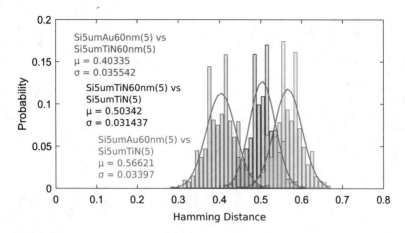

Fig. 4.12 Inter-FHD for five NPs. The underlying spectra are provided in Fig. 4.13

Regarding uniqueness, besides the configurations already covered in Figs. 4.10 and 4.11, Knechtel et al. investigated further peo-PUFs. In Fig. 4.12, the authors provide three inter-FHD distributions for exemplary arrangements of five NPs. The mean values range from 0.4 to 0.56, with reasonably low standard deviations of 0.03. The distributions attest to the potential for strong uniqueness of peo-PUFs. It should be emphasized again that, in reality, considerably more number of NPs will be present. Therefore, the inter-FHD distributions and uniqueness can be expected to improve even further. The transmission plots as well as one arrangement of NPs related to Fig. 4.12 are illustrated in Fig. 4.13. Note that most of the NPs were placed in the middle of the SDR (inset Fig. 4.13b), where the interaction of the propagating photonic mode with NPs is relatively weak. This interaction can be largely enhanced through simple manufacturing means, e.g., by placing a metallic scatterer inside the SDR, to raise the uniqueness of peo-PUFs even further.

4.5   Closing Remarks

PUFs have become increasingly important for applications ranging from key generation and authentication to cryptographic protocols like oblivious transfer and multi-party computation [GAA20]. Hence, it is important to examine their key properties of unpredictability, unclonability, uniqueness, reproducibility, and tamper-resilience and assess their shortcomings against upcoming machine learning-based attacks. This chapter expounded on the problem of limited randomness in CMOS-based PUFs, which can be addressed by adopting emerging technologies that have proved to be more promising with regard to their intrinsic entropy and nonlinearity. This chapter further shed light on some seminal emerging technology-based PUF implementations in the literature and presented a detailed case study

Fig. 4.13 Transmission plots for different peo-PUF tokens with five NPs. The red, blue, and green curves correspond to the setups *Si5umAu60nm(5)*, *Si5umTiN60nm(5)*, and *Si5umTiN(5)*, respectively, as described in Table 4.1. The inset in (**b**) shows the spatial arrangement of NPs for *Si5umTiN60nm(5)*, with NPs labeled as "metal disk"

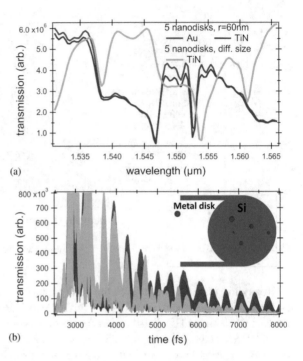

(a)

(b)

on a plasmonics-enhanced optical PUF, which works on the principle of (1) light propagation in a silicon disk resonator and (2) surface plasmon generation from nanoparticles arranged randomly on top of the resonator.

References

[Ata+18] I. Atakhodjaev et al., Investigation of deep learning attacks on nonlinear silicon photonic PUFs, in *Conference on Lasers and Electro-Optics* (2018), FM1G.4. https://doi.org/10.1364/CLEO_QELS.2018.FM1G.4

[Bas+10] L.E. Bassham et al., A statistical test suite for random and pseudorandom number generators for cryptographic applications. Tech. rep. National Institute of Standards and Technology, 2010

[Boz09] S. Bozhevolnyi, *Plasmonic Nanoguides and Circuits* (Pan Stanford, Singapore, 2009)

[BPK19] A. Bojesomo, S. Patnaik, J. Knechtel, *Security Evaluation Framework for peo-PUF*. DfX, NYUAD. 2019. https://github.com/DfX-NYUAD/peo-PUF

[Che+11] W.T. Chen et al., Manipulation of multidimensional plasmonic spectra for information storage. App. Phys. Lett. **98**(17), 171106 (2011). https://doi.org/10.1063/1.3584020

[Che+18] B. Chen et al., Hybrid photon–plasmon coupling and ultrafast control of nanoantennas on a silicon photonic chip. Nano Lett. **18**(1), 610–617 (2018). https://doi.org/10.1021/acs.nanolett.7b04861

[Cui+14] Y. Cui et al., Encoding molecular information in plasmonic nanostructures for anti-counterfeiting applications. Nanoscale **6**, 282–288 (2014). https://doi.org/10.1039/C3NR04375D

[CZZ17] C.H. Chang, Y. Zheng, L. Zhang. A retrospective and a look forward: fifteen years of physical unclonable function advancement. IEEE Circ. Syst. Mag. **17**(3), 32–62 (2017). ISSN: 1531-636X. https://doi.org/10.1109/MCAS.2017.2713305

[GAA20] Y. Gao, S.F. Al-Sarawi, D. Abbott, Physical unclonable functions. Nat. Electron. **3**(2), 81–91 (2020)

[Gan17] F. Ganji, On the learnability of physically unclonable functions. PhD thesis. Technische Universität Berlin, 2017. https://doi.org/10.14279/depositonce-6174

[Gos+17] J. Gosciniak et al., Study of high order plasmonic modes on ceramic nanodisks. Opt. Express **25**(5), 5244–5254 (2017). https://doi.org/10.1364/OE.25.005244

[Gos+19] J. Gosciniak et al., Plasmonic Schottky photodetector with metal stripe embedded into semiconductor and with a CMOS-compatible titanium nitride. Nat. Sci. Rep. **9**(1), 6048 (2019)

[GR19] J. Gosciniak, M. Rasras, High field enhancement between transducer and resonant antenna for application in bit patterned heat-assisted magnetic recording. Opt. Express **27**(6), 8605–8611 (2019)

[Gru+17b] B.C. Grubel et al., Silicon photonic physical unclonable function. Opt. Express **25**(11), 12710–12721 (2017). https://doi.org/10.1364/OE.25.012710

[Gua+18] H. Guan et al., Widely-tunable, narrow-linewidth III-V/silicon hybrid external-cavity laser for coherent communication. Opt. Express **26**(7), 7920–7933 (2018). https://doi.org/10.1364/OE.26.007920

[Her+14] C. Herder et al., Physical unclonable functions and applications: a tutorial. Proc. IEEE **102**(8), 1126–1141 (2014). ISSN: 0018-9219. https://doi.org/10.1109/JPROC.2014.2320516

[Hua+07] H.J. Huang et al., Plasmonic optical properties of a single gold nano-rod. Opt. Express **15**(12), 7132–7139 (2007)

[Kne+19] J. Knechtel et al., Toward physically unclonable functions from plasmonics-enhanced silicon disc resonators. J. Lightwave Tech. **37**(15), 3805–3814 (2019). https://doi.org/10.1109/JLT.2019.2920949

[Kum+13] A. Kumar et al., Dielectric-loaded plasmonic waveguide components: going practical. Laser Photon. Rev. **7**(6), 938–951 (2013)

[KZ12] M. Kauranen, A.V. Zayats, Nonlinear plasmonics. Nat. Photonics **6**, 737–748 (2012). https://doi.org/10.1038/NPHOTON.2012.244

[LF18] K. Li, A.C. Foster, Parametric nonlinear silicon-based photonics. Proc. IEEE **106**(12), 2196–2208 (2018). ISSN: 0018-9219. https://doi.org/10.1109/JPROC.2018.2876515

[LHH18] L. Liu, H. Huang, S. Hu, Lorenz chaotic system-based carbon nanotube physical unclonable functions. Trans. Comput. Aided Des. Integr. Circ. Syst. **37**(7), 1408–1421 (2018). ISSN: 0278-0070. https://doi.org/10.1109/TCAD.2017.2762919

[Liu+18] Y. Liu et al., A combined optimization-theoretic and sidechannel approach for attacking strong physical unclonable functions. Trans. VLSI Syst. **26**(1), 73–81 (2018). ISSN: 1063-8210. https://doi.org/10.1109/TVLSI.2017.2759731

[Mae+15] R. Maes et al., Secure key generation from biased PUFs. Proc. Cryptogr. Hardw. Embed. Syst. (2015). https://doi.org/10.1007/978-3-662-48324-4_26

[MV10] R. Maes, I. Verbauwhede, Physically unclonable functions: a study on the state of the art and future research directions, in *Towards Hardware-Intrinsic Security: Foundations and Practice*, ed. by A.-R. Sadeghi, D. Naccache (Springer, 2010), pp. 3–37. ISBN: 978-3-642-14452-3. https://doi.org/10.1007/978-3-642-14452-3_1

[Nil+18] H. Nili et al., Hardware-intrinsic security primitives enabled by analogue state and nonlinear conductance variations in integrated memristors. Nat. Electron. **1**(3), 197–202 (2018). ISSN: 2520-1131. https://doi.org/10.1038/s41928-018-0039-7

[Nog07] C. Noguez, Surface plasmons on metal nanoparticles: the influence of shape and physical environment. J. Phys. Chem. C **111**(10), 3806–3819 (2007)

[Orc+12] J.S. Orcutt et al., Open foundry platform for high-performance electronic-photonic integration. Opt. Express **20**(11), 12222–12232 (2012). https://doi.org/10.1364/OE.20.012222

[Pap+02] R. Pappu et al., Physical one-way functions. Science   **297**(5589), 2026–2030 (2002). ISSN: 0036-8075. https://doi.org/10.1126/science.1074376. eprint: http://science.sciencemag.org/content/297/5589/2026.full.pdf

[Par+16] K. Park et al., Plasmonic nanowire-enhanced upconversion luminescence for anti-counterfeit devices. Adv. Funct. Mater. **26**(43), 7836–7846 (2016). https://doi.org/10.1002/adfm.201603428. eprint: https://onlinelibrary.wiley.com/doi/pdf/10.1002/adfm.201603428

[Rah+17] F. Rahman et al., Security beyond CMOS: fundamentals, applications, and roadmap. Trans. VLSI Syst. **PP.99**, 1–14 (2017). ISSN: 1063-8210. https://doi.org/10.1109/TVLSI.2017.2742943

[RDK12] U. Rührmair, S. Devadas, F. Koushanfar, Security based on physical unclonability and disorder, in *Introduction to Hardware Security and Trust*, ed. by M. Tehranipoor, C. Wang (Springer, New York, 2012), pp. 65–102. ISBN: 978-1-4419-8080-9. https://doi.org/10.1007/978.1.4419.80809_4

[Ros+13] G.S. Rose et al., Hardware security strategies exploiting nanoelectronic circuits, in *Proc. Asia South Pac. Des. Autom. Conf.* (2013), pp. 368–372. https://doi.org/10.1109/ASPDAC.2013.6509623

[Rüh+13a] U. Rührmair et al., PUF modeling attacks on simulated and silicon data. Trans. Inf. Forens. Sec. **8**(11), 1876–1891 (2013). ISSN: 1556-6013. https://doi.org/10.1109/TIFS.2013.2279798

[Rüh+13b] U. Rührmair et al., Optical PUFs reloaded, in *IACR Crypt. ePrint Arch.* (2013). https://eprint.iacr.org/2013/215

[Sch+10] J.A. Schuller et al., Plasmonics for extreme light concentration and manipulation. Nat. Mater. **9**, 193–204 (2010). https://doi.org/10.1038/nmat2630

[Sei+18] M. Seifried et al., Monolithically integrated CMOS-compatible III-V on silicon lasers. J. Sel. Top. Quant. Electron. **24**(6), 1–9 (2018). ISSN: 1077-260X. https://doi.org/10.1109/JSTQE.2018.2832654

[SKD12] J.A. Scholl, A.L. Koh, J.A. Dionne, Quantum plasmon resonances of individual metallic nanoparticles. Nature **483**(7390), 421 (2012)

[Sor06] R. Soref, The past, present, and future of silicon photonics. J. Sel. Top. Quantum Electron. **12**(6), 1678–1687 (2006). ISSN: 1077-260X. https://doi.org/10.1109/JSTQE.2006.883151

[SPS16] A.F. Smith, P. Patton, S.E. Skrabalak, Plasmonic nanoparticles as a physically unclonable function for responsive anti-counterfeit nanofingerprints. Adv. Funct. Mater. **26**(9), 1315–1321 (2016). https://doi.org/10.1002/adfm.201503989. eprint: https://onlinelibrary.wiley.com/doi/pdf/10.1002/adfm.201503989

[SS17] A.F. Smith, S.E. Skrabalak, Metal nanomaterials for optical anti-counterfeit labels. J. Mater. Chem. C **5**(13), 3207–3215 (2017). https://doi.org/10.1039/C7TC00080D

[Taj17] S. Tajik, On the physical security of physically unclonable functions. PhD thesis. Technische Universität Berlin, 2017. https://doi.org/10.14279/depositonce-6175

[TŠ07] P. Tuyls, B. Škorić, Strong authentication with physical unclonable functions, in *Security, Privacy, and Trust in Modern Data Management*, ed. by M. Petković, W. Jonker (Springer, Berlin, 2007), pp. 133–148. ISBN: 978-3-540-69861-6. https://doi.org/10.1007/978-3-540-69861-6_10

[TŠK07] P. Tuyls, B. Škorić, T. Kevenaar (eds.), *Security with Noisy Data. On Private Biometrics, Secure Key Storage and Anti-Counterfeiting* (Springer, New York, 2007). https://doi.org/10.1007/978-1-84628-984-2

[Wan+13] T. Wang et al., Multi-photon absorption and third-order nonlinearity in silicon at mid-infrared wavelengths. Opt. Express **21**(26), 32192–32198 (2013)

[Wan+15a] C. Wang et al., TSV-based PUF circuit for 3DIC sensor nodes in IoT applications, in *Proc. Electron. Dev. Solid State Circ.* (2015), pp. 313–316. https://doi.org/10.1109/EDSSC.2015.7285113

[WS06] F. Wang, Y.R. Shen, General properties of local plasmons in metal nanostructures. Phys. Rev. Lett. **97**(20), 206806 (2006). https://doi.org/10.1103/PhysRevLett.97.206806

[WYM14] M. Wang, A. Yates, I.L. Markov, Super-PUF: integrating heterogeneous physically unclonable functions, in *Proc. Int. Conf. Comp.-Aided Des.* San Jose, California (2014), pp. 454–461. ISBN: 978-1-4799-6277-8. https://doi.org/10.1109/ICCAD. 2014.7001391

[Xio+16] C. Xiong et al., Monolithic 56 Gbs silicon photonic pulse-amplitude modulation transmitter. Optica **3**(10), 1060–1065 (2016). https://doi.org/10.1364/OPTICA.3.001060

[Zhe+16] Y. Zheng et al., Unclonable plasmonic security labels achieved by shadow-mask-lithography-assisted self-assembly. Adv. Mater. **28**(12), 2330–2336 (2016). https://doi.org/10.1002/adma.201505022. eprint: https://onlinelibrary.wiley.com/doi/pdf/10.1002/adma.201505022

Chapter 5
Intrinsic Entropy for True Random Number Generation

5.1 Chapter Introduction

True Random Number Generators (TRNGs) form an essential and indispensable part of modern security systems in various scenarios, including (1) key- and initial counter value-generation for cryptographic functions, (2) seeding unique device keys for watermarking, (3) authentication in protocols such as the blockchain, and (4) input randomization for physical security measures such as side-channel countermeasures or camouflaging [Neu19]. If the entropy source driving the TRNG is not resilient, the security guarantees concerning these applications may be compromised. Physical attacks such as those based on temperature fluctuations, under/over-volting, and frequency injection [Yan16, MM09] are capable of destroying the entropy source of the TRNG, thus rendering it useless. Hence, there is a need for a robust, reliable, and high-fidelity TRNG for on-chip implementation in security systems. In this chapter, we explore how the intrinsic entropy arising from the complex physical phenomena in emerging devices can be exploited to design such resilient TRNGs.

We first delve into the process of generating random numbers using emerging device technologies, followed by a brief review of prior emerging device-based TRNG implementations. Finally, we present a detailed case study on a spintronics-based TRNG to provide the reader with insights on all aspects of designing a TRNG, from harnessing the entropy source to constructing the sampling circuit, testing its resilience, and benchmarking its performance.

N. Rangarajan et al., *The Next Era in Hardware Security*,
https://doi.org/10.1007/978-3-030-85792-9_5

5.2 Concepts for True Random Number Generation Using Emerging Technologies

Typically, the first step in designing a TRNG is identifying a suitable entropy source, whose randomness can be transduced into a sequence of binary digital bits. In this regard, emerging devices offer a vast array of opportunities to tap into the innate randomness found in nature's physical processes. Quantum sources of entropy, such as vacuum state fluctuations, entangled photons, and Raman scattering, have been harvested to forge TRNGs in the past [Gab+10, Fio+07, Bus+11]. TRNGs exploiting chaos in semiconductor lasers were demonstrated in [UA+08]. The oscillatory and stochastic threshold switching in the insulator-to-metal transition in VO_2 is used to implement a working TRNG model in [Jer+17]. Here, the stochastic threshold switching is enabled by small perturbations in the nanoscale domain structure of VO_2. Wei et al. [Wei+16] exploit the current difference in the $1/f^\beta$ noise of Resistive Random Access Memory (RRAM) to fashion their entropy source. The spatio-temporal differences in the visible spectrum of frequencies is employed to achieve random number generation in [Lee+18].

Broadly, the entropy sources employed in various TRNGs can be classified as follows [Lee+18].

- Electric and electronic noise, including thermal noise, flicker noise, shot noise, and diffusion noise in semiconductor devices.
- Metastability in electronic circuits and natural phenomena, such as ring oscillators, flip flops, and magnetic systems.
- Chaos in Lorenz systems like lasers and optical cavities.
- Physical systems, encompassing photonic, and radioactive decay-based arrangements.

In constructing such emerging device-based TRNGs, there are some considerations to be accounted for.

1. **Purity of entropy source:** Presence of any correlations or bias in the source will adversely affect the quality of randomness. For instance, the TRNG driven by chaos in semiconductor lasers [UA+08] suffers from periodicity and lack of sufficient intrinsic randomness.
2. **Difficulty of extracting entropy:** Significant challenges in handling and extraction of the entropy source, e.g., in radioactive decay-based TRNGs [Par+20], often make it impractical.
3. **Complexity of post-processing required:** Most of the quantum TRNGs [Gab+10, Bus+11] require complex photo-detection and enormous post-processing, resulting in unfeasible overheads.
4. **Scalability:** Optical source-based TRNGs [Fio+07] often suffer from scalability issues due to size of components involved.
5. **Ease of CMOS integration:** The underlying technology must be compatible with the existing CMOS framework, for economic viability.

5.3 Review of Selected Emerging Technologies and Prior Art

In this section, we briefly review prior TRNGs, which leverage various classes of emerging devices and phenomena. The purpose of this survey is to get a better perspective of how emerging devices and materials can be moulded into a physical entropy generating contraption.

5.3.1 Ferroelectric Field Effect Transistor

A ferroelectric field effect transistor (FeFET) is a modification of the conventional MOSFET, wherein a ferroelectric material is sandwiched between the gate electrode and the channel [KKD20]. The application of a gate voltage causes polarization of the ferroelectric domains, thus imparting memory to the device. The direction of this polarization (up or down) can either support or impede the formation of the inversion channel, thereby affecting the threshold voltage.

This tunability of the inversion channel, by switching the polarization charge in the ferroelectric layer, is utilized by Mulaosmanovic et al. [MMS17] to build a TRNG. The operation of their TRNG is as follows. The circuit is built using a single FeFET device, with a polysilicon/TiN/HfO$_2$/SiON gate stack. The application of a positive gate voltage orients the ferroelectric polarization downward and sets the device into low-V_T state, whereas a negative gate voltage causes upward polarization and resets the device to its high-V_T state (Fig. 5.1). On reducing the dimensions of the FeFET device, such that the channel length is on the order of the ferroelectric domain size, the switching process from one state to the other becomes extremely sharp and unpredictable. This stochastic switching occurs close to the ferroelectric coercive voltage, wherein the pulse width at which the abrupt switching occurs exhibits variability. The randomness in the ferroelectric switching is attributed to the variability in the nucleation-driven polarization reversal of the domains. By tuning the pulse width for a 50% probability of switching from high- to low-V_T state, equiprobable random numbers can be generated. The high- or low-V_T states are converted to binary digits by sampling the drain current levels and then using a comparator setup.

The FeFET-based TRNG proposed in this work can be highly scalable owing to its 1-transistor structure, and offers advantages like low power operation and CMOS compatibility. However, it would be prone to process and temperature variability since the margins for the 50% switching threshold are limited.

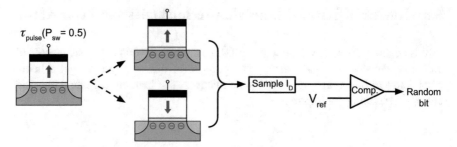

Fig. 5.1 Schematic of the ferroelectric switching based-TRNG. The gate pulse width is tuned to achieve 50% switching probability, thereby resulting in a random V_T state. After the switching process, the drain current at that final V_T state is sampled and converted to a digital bit

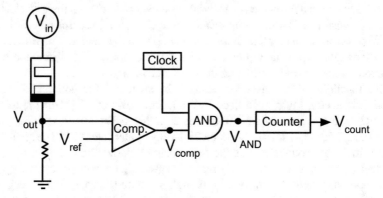

Fig. 5.2 Circuit construction of the diffusive memristor-based TRNG. A voltage divider arrangement with a series resistor is used to read the output voltage. The sampling circuit consists of a comparator, an AND gate, and a counter

5.3.2   Diffusive Memristor

Hao Jiang et al. demonstrated in [Jia+17] a novel TRNG based on the stochastic switching behavior in diffusive memristors. Their Ag:SiO$_2$ diffusive memristor is a volatile device, which functions on the basis of the stochastic diffusion dynamics of metal atoms in the memristive channel. The application of a voltage pulse turns the device ON and switches it to a low-resistance state. However, the delay incurred in this process is random and uncontrollable. The device is turned OFF when the bias is removed, and the channel relaxes back to the high-resistance state. The innate stochasticity prevalent in the switching delay of the memristor is exploited as the entropy source.

The TRNG circuit (Fig. 5.2) is area-efficient and is composed of (1) the diffusive memristor for the entropy, and (2) a sampling circuit constructed with a comparator, AND gate, and a counter. The memristive device itself consists of a Pt/Ag/Ag:SiO$_2$/Pt stack, where a 5 nm Ag reservoir layer is placed between the

Ag:SiO$_2$ channel and the top electrode. The working of their TRNG circuit is as follows. A voltage bias V_{in} above the threshold voltage (0.5 V) causes Ag atoms to randomly detach from the Ag reservoir layer and diffuse toward the bottom Pt electrode. After arbitrary time, enough Ag atoms have detached and migrated to form a conductive channel, thus switching the device ON. Since the detaching process of Ag atoms is a random phenomenon, the time taken for the onset of conduction is random as well.

As the device turns ON, the output voltage (V_{out}) between the memristor and series resistor rises. When this output voltage increases beyond a preset reference, the comparator output (V_{comp}) goes high and stays high until the input voltage pulse V_{in} to the memristor is cut-off. A random delay time to reach the ON state ensures that the pulse width of V_{comp} is random. Hence, AND-ing this V_{comp} with a (higher frequency) clock signal produces a waveform with arbitrary number of clock cycles encompassed within the random pulse width of V_{comp}. The AND-ed signal V_{AND} is sent to a counter, whose output V_{count} flips at each rising edge of V_{AND}, and settles at "1" or "0" when V_{in} goes low. The bit on which the counter settles is random due to the random pulse width of V_{comp}, which in turn causes a random number of clock pulses to be sent to the counter.

The entropy source in this TRNG is shown to be sufficiently robust even at elevated temperatures, and requires minimal post-processing. The normally OFF and volatile nature of the diffusive memristor implies that no reset stage is required, hence reducing the energy consumption. However, this particular TRNG implementation suffers from poor bitrate (maximum ~300 Kbps) and endurance ($\sim 10^7$ cycles).

5.3.3   Spin Dice

The spin dice, introduced by Fukushima et al. [Fuk+14], was one of the first spintronics-based scalable TRNGs, built by extracting the stochastic nature of spin-transfer torque (STT) switching. Here, the STT mechanism, [RS08, LZ03, XZS05], predicted by J. Slonczewski and L. Berger, is used as the fundamental mechanism to control the magnetization state of the nanomagnet.

The flipping of magnetization state is inherently probabilistic due to (1) the presence of thermal noise in nanomagnets and (2) the existence of two equivalent basins of attraction in the energy landscape of nanomagnets that make the dynamical evolution of magnetization extremely sensitive to its initial state. As a consequence, a nanomagnet brought to an unstable initial state will relax to a final magnetization orientation that appears to be unpredictable. Depending on the relative polarizations of the input spin current and the magnetization vector, two cases of magnetization reversal can occur: (1) damping switching [Ber+03] and (2) precessional switching [dAq05, Liu+10]. In the case of damping switching, the spin current entering the nanomagnet is polarized in the plane of the nanomagnet. The magnetization oscillates around the local minimum with a slowly increasing amplitude, and at

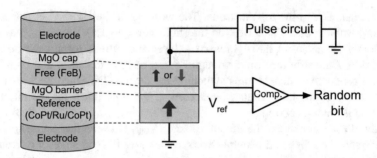

Fig. 5.3 Perpendicularly magnetized MTJ (p-MTJ) stack and circuit schematic used to construct spin dice

some point the projection on the easy axis abruptly switches to the opposite value. In precessional switching, the magnetization is pushed along the path of the steepest ascent, and a very short duration of the spin current can induce switching.

The setup in [Fuk+14] consists of top-free perpendicularly magnetized Magnetic Tunnel Junctions (p-MTJs), constructed using a synthetic antiferromagnetic bottom reference layer. Such p-MTJs have been shown to exhibit a wide magnetic field range for the bistable states around zero magnetic field and low switching current densities. Their p-MTJ stack is composed of a 2 nm FeB free layer, a 1 nm MgO barrier, and a CoPt/Ru/CoPt reference layer. The stack is fabricated as a nano-pillar with a cross-section of $70 \times 200\,\text{nm}^2$ and a magnetoresistance ratio of 100%. Figure 5.3 illustrates the spin dice p-MTJ stack as well as the circuit schematic for generating random bits. The working of the spin dice is as follows. A current pulse from the pulse circuit is applied to the p-MTJ to perturb it from its initial state. The pulse width and amplitude of this current pulse are carefully tuned to achieve a switching probability of 50% for the considered free layer. The final state of the free ferromagnet is obtained by measuring the resistance of the p-MTJ stack, which is converted to a digital bit using a comparator circuit. Further details on the STT switching of MTJ circuits, and their subsequent read-out and digital sampling are presented in Sect. 5.4.2.

Although the spin dice offers a stable operation with high integration density, it is prone to issues arising from temperature fluctuations and process variations. The damping switching mechanism inherently has a very narrow bistable region, and small deviations in the operating temperature or the input current pulse can change the switching probability from the required 50%. This can significantly deteriorate the randomness of the TRNG. Hence, the spin dice requires temperature compensation and additional post-processing to high quality of randomness.

5.4   Case Study: Precessional Nanomagnet Switching for TRNG

Having explored the operation of prior emerging device-based TRNG implementations, we now look at another spintronics TRNG in much more detail. Through this case study, we shed light on the process of designing an emerging device-based TRNG from scratch, including (1) the selection and evaluation of a viable entropy source, (2) moulding the entropy source into a system-level TRNG implementation, and finally (3) testing its randomness and performance.

The spintronics-based TRNG in [RPR17] leverages the inherent stochasticity of the precessional magnetization dynamics in thin-film nanomagnets. Here, similar to spin dice, the STT mechanism is used to induce nanomagnet switching. The time-dependent evolution of the magnetization in a thin-film nanomagnet, under the influence of STT, is obtained using the stochastic Landau–Lifshitz–Gilbert–Slonczewski (s-LLGS) equation [ARR16]. Readers are referred to Appendix 1 for further details on the s-LLGS dynamics.

Precessional switching is excited in the multilayer structure shown in Fig. 5.4a, where the polarizing magnet is orthogonal to the free layer (implementation details presented in Sect. 4.3). Sample magnetization reversal trajectories corresponding to damping and precessional switching mechanisms obtained by solving the s-LLGS equation are shown in Fig. 5.4b,c.

Note here that the precessional switching mechanism is more prone to stochasticity than damping switching. In the case of damping switching, the magnetization traverses from one equilibrium point to the other under the effect of STT, through the saddle point, without ever leaving the plane of the thin-film nanomagnet (x-y plane). However, in precessional switching, the magnetization is assisted out of the plane by the STT, which is a high energy region in the energy landscape of the nanomagnet. This causes precessional switching to be inherently more chaotic, and this chaos coupled with the multistability, fine entanglement of the basins of attraction, and

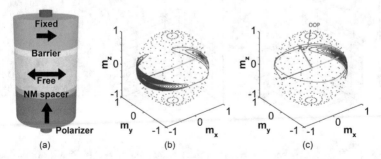

Fig. 5.4 (**a**) MTJ stack arrangement used in the precessional switching-based TRNG circuit. (**b**) and (**c**) are sample magnetization trajectories for damping and precessional switching, respectively. In precessional switching, the magnetization is pushed out-of-plane (OOP) before relaxing to an in-plane stable orientation ($\pm \hat{m}_x$)

extreme sensitivity to the initial conditions results in a probabilistic switching mechanism even in the absence of thermal fluctuations [dAq+15, BSM13]. Hence, as compared to the spin dice described in Sect. 5.3.3, which operates on the damping switching mechanism [Fuk+14], the precessional switching-based TRNG has negligible dependence on thermal effects and can be operated over a wide range of temperature as well as the strength of the STT effect.

5.4.1 Entropy Source

Solving the s-LLGS equation with no applied field or spin current results in six stationary points: two stable minimum energy equilibrium points at $\mathbf{m} = \pm\hat{\mathbf{x}}$, two saddle points at $\mathbf{m} = \pm\hat{\mathbf{y}}$, and two unstable maximum energy points at $\mathbf{m} = \pm\hat{\mathbf{z}}$. The energy portrait of precessional switching projected on the m_x-m_y plane is shown in Fig. 5.5a. In this figure, the shaded regions (low energy) are the regions where the magnetic free energy is below the energy of the saddle points, while in the elliptical white (high energy) region, the energy is above the energy of the saddle points [BMS03]. The energy portrait is also demonstrated by the simulation of magnetic relaxations when the magnetization vector is initialized at different directions in Fig. 5.5b. The red dots inside the ellipse $(1 + D)m_x^2 + m_y^2 = 1$ correspond to initializations that relax toward one equilibrium point, and the green dots correspond to relaxation toward the other.

The precessional switching mechanism of a nanomagnet subject to the STT effect is a two step process. In the first step, the magnetization undergoes out-of-plane

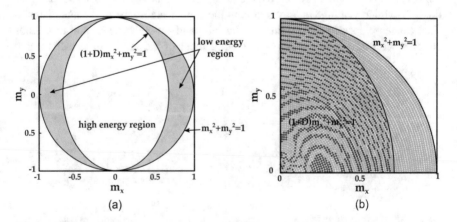

Fig. 5.5 (**a**) High energy and low energy regions on m_x-m_y plane. (**b**) Result of relaxation of magnetization for different initial magnetization states on m_x-m_y plane. Green dots and red dots are the relaxed states corresponding to $\mathbf{m} = \hat{\mathbf{x}}$ and $\mathbf{m} = -\hat{\mathbf{x}}$, respectively. Here, the ratio of the anisotropy field and the saturation magnetization is unity, and the Gilbert damping constant $\alpha = 0.01$

precession due to the spin current polarized in $\hat{z}$ direction, and enters the high energy region. In the second step, the spin current is switched off after the magnetization has completed a quarter precession around $\hat{z}$, and the magnetization undergoes relaxation oscillations toward one of the equilibrium points. Since dissipative effects result in a decrease of the magnetic free energy, the time evolution of relaxations within any shaded region inevitably leads to the equilibrium point inside that region. But for our case, the magnetization relaxation starts in the white region, so depending on the initial conditions it may settle to one of the two stable points in the shaded regions. The number of green and red dots inside the high energy region in Fig. 5.5b is roughly the same, resulting in a 50% probability of relaxation to either stable state.

5.4.2 Implementation

The precessional switching-based TRNG device is constructed from an MTJ arrangement as shown in Fig. 5.4a. The MTJ stack consists of a magnetic polarizer whose equilibrium magnetization vector is oriented in the perpendicular direction, while the free and fixed magnetic layers have their equilibrium magnetization vectors oriented in the plane of the film. A non-magnetic copper spacer separates the free layer from the polarizer, while an MgO tunnel barrier is used to separate the free and fixed layers. The free layer is a CoFeB ferromagnet while the fixed layer is composed of a CoFeB/Ru/CoFeB stack. The material and geometrical parameters of the MTJ stack used for simulations in this case study are listed in Table 5.1.

On applying a bias to the MTJ stack, a current flows through the bottom layer, which polarizes the spins in the incoming current and orients them orthogonal to the magnetization of the free ferromagnet. This polarized spin current then causes the free ferromagnet magnetization to precess out-of-plane and reach the bistable region. Here, the magnetization vector sees two equivalent basins of attraction on both sides, and could switch to either configuration with a 50% probability, as shown in Fig. 5.6.

Depending on whether the final magnetization of the free ferromagnet is parallel or anti-parallel with respect to the top reference layer magnetization, the

Table 5.1 Geometrical and material parameters of the MTJ stack

Parameter	
Cross-section of free layer	$75.6 \times 37.8 \, nm^2$
Thickness of Cu spacer	2 nm
Thickness of free layer	1.5 nm
Anisotropy field (H_k) of free layer	$1.92 \times 10^5 \, A/m$ [Ike+10]
Saturation magnetization (M_s) of free layer	$1.92 \times 10^5 \, A/m$
Thickness of MgO barrier	1 nm

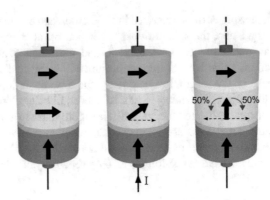

Fig. 5.6 The magnetization of the central free layer (green), taken out-of-plane into the bistable region through the application of electric current, can switch to either of the two equilibria with an equal probability

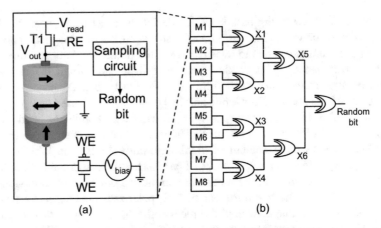

Fig. 5.7 (**a**) TRNG circuit. RE and WE are the read and write enable signals, respectively. The sampling circuit consists of a sense amplifier and a D flip-flop (**b**) XOR ladder of TRNG units. Each of the M's is an individual TRNG circuit

device enters its low-resistance (LR) or high-resistance (HR) mode of operation, respectively. Now, a read bias (V_{read}) applied on the potential divider configuration of transistor T1 and the MTJ will result in a high or low voltage level, respectively, at the output node (V_{out}) connected to the sampling circuit. This output voltage is compared with a reference voltage, and the corresponding bit (1 or 0) is produced by the sampling circuit (Fig. 5.7a). The reference voltage is chosen to lie between the high and low voltage levels generated from the MTJ stack. This provides sufficient low and high noise margins for signal detection even in the presence of mismatch in the differential pair transistors of the sense amplifier in the sampling circuit.

It is customary to XOR the outputs of two TRNG units with one another to augment the Shannon entropy, since XOR has the property of entropy accumulation

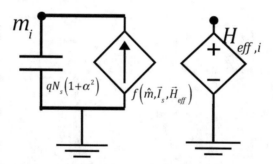

Fig. 5.8 SPICE implementation of the nanomagnet in one specific direction. Here, m_i is treated as a voltage, while the magnetic fields acting on the nanomagnet are represented through a voltage source H_{eff}. The effect of STT on the nanomagnet is modeled using the current source, f. See [Bon+14] for implementation details

[Mat93] (Fig. 5.7b). This ensures the swellness of the random bits produced and also makes the TRNG more robust and less vulnerable to corruption of any single entropy source. Even if any of the entropy sources fail, the XORed output will still be random enough to be used in applications requiring a high quality of randomness. Note that the binary sequences generated by the individual TRNG units are uncorrelated, since the entropy sources (driven by thermal noise) for the individual units are uncorrelated themselves.

To simulate the circuit-level behavior of the precessional switching-based TRNG circuit, physics-based models of the ferromagnets, interfaces, and the complete MTJ device are developed in SPICE. The SPICE circuit implementation of a nanomagnet subject to thermal effects and spin torque is shown in Fig. 5.8. This circuit models the magnetization of the nanomagnet in one particular direction (x, y, or z), and the other two magnetization directions are implemented similarly. The magnetization m_i is modeled as the node voltage of a capacitor, and the effective field inside the nanomagnet is represented as a dependent current source. Detailed circuit derivations can be found in prior works [Bon+14, MNY12]. The MTJ SPICE model is integrated with the CMOS circuitry in 45 nm technology to obtain the full TRNG circuit. BSIM4 Level 54 models are used for the transistors in the circuit. Convergence to the implicit midpoint Stratonovich solution of the s-LLGS equation [ARR16] is obtained through the trapezoidal SPICE solver, with a minute time step (1 fs). Further, the layout for the circuit is constructed to obtain the post-layout area.

5.4.3 Benchmarking Randomness and Performance

After determining the entropy source and designing the TRNG circuit, it is crucial to test the quality of random numbers generated from the circuit. A simple preliminary test to evaluate whether the TRNG meets the expected standards is to analyze

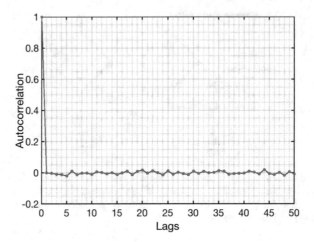

Fig. 5.9 Average autocorrelation of 10 bit streams (10,000 bits long) generated by the proposed TRNG at 300 K, at various lags

the autocorrelation of the bit streams generated. For the precessional switching-based TRNG in Sect. 5.4.2, the autocorrelation of samples of 10,000 bits extracted from a large bit stream (one million bits), churned out by the TRNG at 300 K, is investigated. Figure 5.9, which highlights the results of this analysis, shows that the average autocorrelation of the samples at various lags (ranging from 1 to 50) is negligible. This preliminary study gives a sense of whether the TRNG is indeed functioning as expected, with further comprehensive NIST randomness tests to follow.

5.4.3.1   NIST Tests for Randomness

The NIST's SP 800-22 statistical test suite for the validation of RNGs [Ruk+01] is used to test bit streams of a million bits. This test suite comprises 15 different tests to evaluate various aspects of randomness for the TRNG considered (See Appendix 2). Each of these tests has an associated figure-of-merit called a p-value, whose expressions can be found in [Ruk+01]. The bit streams to be evaluated are generated from a CUDA-C LLGS solver, which models a nanomagnet switching under the action of STT through the precessional switching mechanism. Further, this test is performed for bit streams generated from all process and temperature corners to appraise the PVT performance of the TRNG. A total of 10 binary sequences are tested for each scenario. The results from these tests for nominal process and $\pm 10\%$ process variability at 300, 200, and 400 K are shown in Fig. 5.10, highlighting the fact that the device operation is insensitive to variations around the nominal process conditions. Here, the $\pm 10\%$ process variation is considered in the thickness of the free ferromagnet. For a thin-film nanomagnet, variations in the thickness would

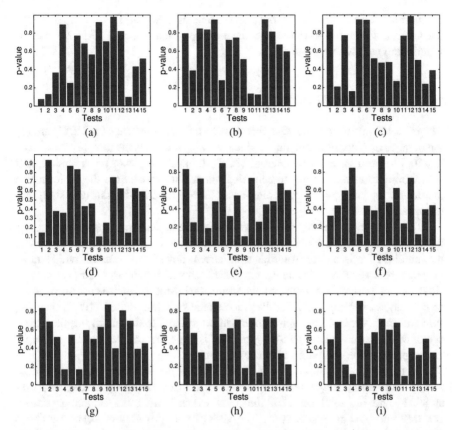

Fig. 5.10 NIST test suite results for the TRNG at 300 K for different process corners. The individual tests are (1) Frequency (monobit), (2) Block frequency, (3) Runs, (4) Longest runs of 1's, (5) Binary matrix rank, (6) DFT, (7) Non-overlapping template, (8) Overlapping template, (9) Maurer's universal, (10) Linear complexity, (11) Serial, (12) Approximate entropy, (13) Cumulative sums, (14) Random excursions, and (15) Random excursions variant. (**a**) 300 K, nominal process. (**b**) 300 K, +10% process. (**c**) 300 K, −10% process. (**d**) 200 K, nominal process. (**e**) 200 K, +10% process. (**f**) 200 K, −10% process. (**g**) 400 K, nominal process. (**h**) 400 K, +10% process. (**i**) 400 K, −10% process

represent the worst case scenario since this is the dimension that has the most impact on the energetics of the magnet, as compared to variability in the length or width, or in the CMOS process. The generated bit streams (10/10 proportion of sequences) pass all the NIST tests for the various process and temperature corners, where a pass is deemed to be a p-value ≥ 0.01, implying 99% confidence levels. For the tests which produce multiple p-values, although individually all of them are above 0.01, the plots in Fig. 5.10 only show the average p-value. Hence, the precessional switching-based TRNG is very robust and reliable, and can operate within a temperature range of 200–400 K and process variations of +10% to −10% without any distortions in the entropy. The entropy source itself is not dependent on thermal effects, and could operate outside of 200–400 K as well; however, the analysis is limited to this range since any physical on-chip TRNG would be functioning well within these limits.

5.4.3.2 Performance Metrics

The most significant metrics that characterize the performance of a TRNG are (1) the number of random bits it can produce in unit time or its bit-rate, (2) the energy required to produce a single random bit, (3) its average power consumption, and (4) its on-chip area for a given technology node. In this section, the performance metrics for the precessional switching-based TRNG described in this case study, are quantified.

In the damping switching mechanism, the injected spin current has a polarization that is collinear to the initial magnetization and easy axis of the ferromagnet. Hence, the initial STT originating from this spin current is very small, and it builds up as the angle between the spins and the magnetization vector increases. This results in a sluggish and gradual switching and, therefore, the bit-rate of such a device operating on damping switching is low. For the precessional switching-based TRNG, since the spin current is applied with spins polarized orthogonal to the initial magnetization vector of the ferromagnet, the STT generated is large; hence, the switching is rapid as compared to the damping switching mechanism.

The temporal evolution of the magnetization of the free ferromagnet is shown in Fig. 5.11. The average switching delay of the free ferromagnet magnetization at 300 K and the total delay of the TRNG device after the sampling process are 3.29 and 3.41 ns, respectively. The CMOS post-processing takes only about 3% of the total delay, while the main limitation in performance stems from the nanomagnet dynamics. The various performance metrics of the TRNG at different temperatures are listed in Table 5.2. The obtained bitrate of 293 Mbps (at 300 K) is significantly higher than that of the Spin Dice [Fuk+14] (0.6 Mbps)

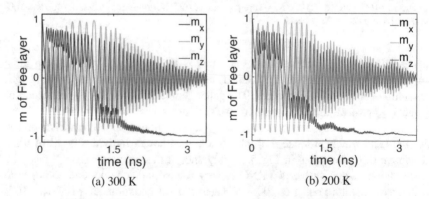

Fig. 5.11 Magnetization reversal of $\hat{m}_x$ of the free nanomagnet at 300 and 200 K. Also shown are the magnetization vectors $\hat{m}_y$ and $\hat{m}_z$, which undergo precessional oscillations

Table 5.2 Performance metrics of the precessional switching-based TRNG at various temperatures. The temperature is varied only for the thermal field of the MTJ free ferromagnet and not for the CMOS circuit. Hence for a fixed bias, the power does not vary with temperature

	200 K	300 K	400 K
Bitrate	303 Mbps	293 Mbps	279 Mbps
Energy-per-bit	338 fJ	351 fJ	382 fJ
Power	136.3 μW		
Area	12.07 μm²		

and the metastable ring oscillator-based TRNG [Vas+08] (140 Mbps). Further, the considered device is inherently stochastic due to the precessional dynamics and, hence, requires minimal post-processing after the random bit is generated. In contrast, the damping switching-based designs [Fuk+14] require additional post-processing to ensure high quality of randomness due to their small operating range (current bias) and susceptibility to process and temperature variations on the chip.

5.5 Closing Remarks

TRNGs are a crucial aspect of modern computing and communication systems, and form the cornerstone for numerous security and cryptographic solutions. The design of an efficient, robust, and reliable on-chip TRNG is vital for secure processor architectures and security chips. This chapter introduces the process of constructing such a TRNG from emerging device-based entropy sources. The types of entropy sources and the factors involved in choosing a suitable source are first identified. This is followed by a brief review of seminal emerging device-based TRNGs and a case study detailing the stepwise implementation of a spintronics TRNG. The insights gained from this chapter are intended to provide the reader with an understanding of the intricacies of designing a TRNG system as well as a notion of what constitutes a good TRNG implementation.

Appendix 1

The dynamics and performance of the majority of spin-based devices are modeled using the stochastic Landau–Lifshitz–Gilbert–Slonczewski (s-LLGS) equation. This equation describes the temporal evolution of the magnetization vector, M, of a monodomain nanomagnet under the effects of magnetic fields, STT, and thermal noise [Slo96, RS08, SZ02]. Mathematically, the s-LLGS equation is given as

$$\frac{dM}{dt} = -\gamma \mu_0 (M \times H_{\text{eff}}) + \frac{\alpha}{M_s}\left(M \times \frac{dM}{dt}\right) - \frac{M \times (M \times I_s)}{q N_s M_s}, \tag{5.1}$$

where H_{eff} is the effective magnetic field experienced by the nanomagnet, α is the dimensionless Gilbert damping constant, I_s is the applied spin current, q is the elementary charge, and other quantities are as defined previously.

The first term on the right hand side (RHS) in Eq. (5.1) is the conservative precessional torque that governs the precession of the magnetization vector around the effective field acting on the nanomagnet. This effective field comprises the magnetocrystalline anisotropy field, the shape anisotropy field, and the external applied field. A Langevin field $h_r = h_x\hat{x} + h_y\hat{y} + h_z\hat{z}$, representing Gaussian white noise, is added into the effective field in the s-LLGS equation to model thermal noise. The second term on the RHS in (5.1) is the Gilbert damping torque, which is responsible for damping the precessions of the magnetization vector and eventually relaxing it to one of its stable states [dAq05]. The final term on the RHS in Eq. (5.1) is the Slonczewski spin torque arising from the deposition of spin angular momentum by the itinerant electrons of the spin-polarized current. For simplicity of analysis, Eq. (5.1) is often transformed into its dimensionless form, expressed as

$$\frac{dm}{dt} = -m \times h_{eff} + \alpha\left(m \times \frac{dm}{dt}\right) - m \times (m \times i_s), \qquad (5.2)$$

where we have the normalized quantities $m = \frac{M}{M_s}$, $h_{eff} = \frac{H_{eff}}{M_s}$, and $i_s = \frac{I_s}{I}$. Here, the scaling factor, I, for spin current is defined as $I = q\gamma\mu_0 M_s N_s$. The time scale is normalized using the factor $(\gamma\mu_0 M_s)^{-1}$. The advantages of the normalized equation Eq. (5.2) over Eq. (5.1) are: (a) it is easier to deal with normalized quantities in terms of numerical complexity, and (b) normalized entities are mathematically well behaved under the application of a numerical scheme. The explicit form of Eq. (5.2) obtained by decoupling dm/dt is given as

$$\frac{dm}{dt} = -\frac{1}{1+\alpha^2}[m \times h_{eff} + (m \times (m \times i_s))$$
$$+ \alpha\left(m \times (m \times h_{eff})\right) - \alpha(m \times i_s)]. \qquad (5.3)$$

To model the thermal field in s-LLGS, it in expressed in terms of the Wiener process as $H_T(t)dt = vdW(t)$ [Aqu+06], where $W(t)$ is the Wiener process, and $v = \sqrt{\frac{2\alpha K_b T}{\mu_0 M_s^2 V}}$ [Sun06, MNY12]. Here, $K_b T$ is the thermal energy. The statistical properties of this thermal field discussed by Brown and Kubo are given as [Bro63, KH70]

(1) The mean thermal field: $\langle H_{T,i}(t)\rangle = 0$,
(2) The correlation between the components of $H_T(t)$ defined over a time interval τ,

$$\langle H_{T,i}(t)H_{T,j}(t+\tau)\rangle = \frac{2K_b T\alpha}{\gamma\mu_0^2 M_s V}\delta_{ij}\delta(\tau), \qquad (5.4)$$

where δ_{ij} is the Kronecker delta function. To simulate the thermal effects numerically, the model is discretized in time

$$H_T(t)\Delta t = \nu \Delta W(t), \tag{5.5}$$

where $\Delta W(t) = W(t+\Delta t) - W(t)$. The normalized standard deviation of the thermal field is given by

$$\sigma = \sqrt{\frac{2\alpha K_b T}{\mu_0 M_s^2 V}} \sqrt{\frac{\Delta t'}{\gamma \mu_0 M_s}}, \tag{5.6}$$

where Δt is the time step of the numerical method used and $t' = (\gamma \mu_0 M_s)t$.

We then have

$$h_T(t)\Delta t' = \sigma \xi_t, \tag{5.7}$$

where the normalized thermal field $h_T = H_T/M_s$, and $\xi_t \sim \mathcal{N}(0,1)$ is a standard Gaussian vector.

Now, the total normalized effective field is given as

$$\begin{aligned}
h_{\text{eff}} = \frac{H_{\text{eff}}}{M_s} &= \frac{-1}{\mu_0 M_s V} \nabla_M E_{\text{total}}(M) + h_T \\
&= h_{\text{app}} + \frac{H_k}{M_s}(\hat{n} \cdot m)\hat{n} - \sum_i N_i m_i + h_T,
\end{aligned} \tag{5.8}$$

where $h_{\text{app}} = \frac{H_{\text{app}}}{M_s}, m_i = \frac{M_i}{M_s}$, and ∇_M is the gradient with respect to the magnetization M.

Appendix 2

A brief description of the various tests encompassed in the NIST SP 800-22 statistical test suite [Ruk+01] is given below.

1. **Frequency (Monobit) Test:** Evaluates proportion of ones and zeroes in the entire sequence.
2. **Frequency Test within a Block:** Divides entire sequence into n-bit blocks and then evaluates proportion of ones within each n-bit block.
3. **Runs Test:** Evaluates the number of uninterrupted runs of identical bits in the sequence.
4. **Test for the Longest-Run-of-Ones in a Block:** Determines the longest uninterrupted sequence of ones in n-bit blocks.

5. **Binary Matrix Rank Test:** Constructs disjoint sub-matrices of the entire sequence and then evaluates their rank.
6. **Discrete Fourier Transform (DFT) Test:** Detects periodic features and peaks in the DFT spectrum of the sequence.
7. **Non-overlapping Template Matching Test:** Matches the sequence against m-bit target string templates in a non-overlapping fashion and returns the number of such occurrences.
8. **Overlapping Template Matching Test:** Matches the sequence against m-bit target string templates in an overlapping fashion and returns the number of such occurrences.
9. **Maurer's Universal Statistical Test:** Identifies similar patterns in the sequence and then evaluates the number of bits between such matching patterns.
10. **Linear Complexity Test:** Calculates the length of a linear feedback shift register for the sequence.
11. **Serial Test:** Determines the frequency of all possible overlapping n-bit patterns in the sequence.
12. **Approximate Entropy Test:** Determines the frequency of all possible overlapping n-bit and (n+1)-bit patterns in the sequence and compares them against statistics for an ideal random sequence.
13. **Cumulative Sums Test:** Converts the sequence of ones and zeroes to $(1, -1)$ and then calculates the maximum excursion (from 0) of a random walk defined by the cumulative sum of the new sequence.
14. **Random Excursions Test:** Evaluates the number of states having K visits in the cumulative sum random walk defined in the previous test.
15. **Random Excursions Variant Test:** Evaluates the number of visits to various states in the cumulative sum random walk defined in the previous tests.

References

[Aqu+06] M. d'Aquino et al., Midpoint numerical technique for stochastic Landau-Lifshitz-Gilbert dynamics. J. Appl. Phys. **99**(8), 08B905 (2006)

[ARR16] S. Ament, N. Rangarajan, S. Rakheja, A practical guide to solving the stochastic Landau-Lifshitz-Gilbert-Slonczewski equation for macrospin dynamics. Preprint. arXiv:1607.04596 (2016)

[Ber+03] G. Bertotti et al., Comparison of analytical solutions of Landau–Lifshitz equation for "damping" and "precessional" switchings. J. Appl. Phys. **93**(10), 6811–6813 (2003)

[BMS03] G. Bertotti, I.D. Mayergoyz, C. Serpico, Critical fields and pulse durations for precessional switching of thin magnetic films. IEEE Trans. Magn. **39**(5), 2504–2506 (2003)

[Bon+14] P. Bonhomme et al., Circuit simulation of magnetization dynamics and spin transport. IEEE Trans. Electron Dev. **61**(5), 1553–1560 (2014)

[Bro63] W.F. Brown Jr, Thermal fluctuations of a single-domain particle. J. Appl. Phys. **34**(4), 1319–1320 (1963)

[BSM13] G. Bertotti, C. Serpico, I.D. Mayergoyz, Probabilistic aspects of magnetization relaxation in single-domain nanomagnets. Phys. Rev. Lett. **110**(14), 147205 (2013)

[Bus+11] P.J. Bustard et al., Quantum random bit generation using stimulated Raman scattering. Opt. Exp. **19**(25), 25173–25180 (2011)

[dAq+15] M. d'Aquino et al., Analysis of reliable sub-ns spin-torque switching under transverse bias magnetic fields. J. Appl. Phys. **117**(17), 17B716 (2015)

[dAq05] M. d'Aquino, Nonlinear magnetization dynamics in thin-films and nanoparticles. PhD thesis. Università degli Studi di Napoli Federico II, 2005

[Fio+07] M. Fiorentino et al., Secure self-calibrating quantum random-bit generator. Phys. Rev. A **75**(3), 032334 (2007)

[Fuk+14] A. Fukushima et al., Spin dice: A scalable truly random number generator based on spintronics. Appl. Phys. Exp. **7**(8), 083001 (2014)

[Gab+10] C. Gabriel et al., A generator for unique quantum random numbers based on vacuum states. Nat. Photonics **4**(10), 711–715 (2010)

[Ike+10] S. Ikeda et al., A perpendicular-anisotropy CoFeB–MgO magnetic tunnel junction. Nat. Mater. **9**(9), 721–724 (2010)

[Jer+17] M. Jerry et al., Stochastic insulator-to-metal phase transition-based true random number generator. IEEE Electron Dev. Lett. **39**(1), 139–142 (2017)

[Jia+17] H. Jiang et al., A novel true random number generator based on a stochastic diffusive memristor. Nat. Commun. **8**(1), 1–9 (2017)

[KH70] R. Kubo, N. Hashitsume, Brownian motion of spins. Prog. Theor. Phys. Suppl. **46**, 210–220 (1970)

[KKD20] A.I. Khan, A. Keshavarzi, S. Datta, The future of ferroelectric field-effect transistor technology. Nat. Electron. **3**(10), 588–597 (2020)

[Lee+18] K. Lee et al., TRNG (True Random Number Generator) method using visible spectrum for secure communication on 5G network. IEEE Access **6**, 12838–12847 (2018)

[Liu+10] H. Liu et al., Ultrafast switching in magnetic tunnel junction based orthogonal spin transfer devices. Appl. Phys. Lett. **97**(24), 242510 (2010)

[LZ03] Z. Li, S. Zhang, Magnetization dynamics with a spin-transfer torque. Phys. Rev. B **68**(2), 024404 (2003)

[Mat93] M. Matsui, Linear cryptanalysis method for DES cipher, in *Workshop on the Theory and Application of of Cryptographic Techniques* (Springer, Berlin, 1993), pp. 386–397

[MM09] A.T. Markettos, S.W. Moore, The frequency injection attack on ring-oscillator-based true random number generators, in *International Workshop on Cryptographic Hardware and Embedded Systems* (Springer, Berlin, 2009), pp. 317–331

[MMS17] H. Mulaosmanovic, T. Mikolajick, S. Slesazeck, Random number generation based on ferroelectric switching. IEEE Electron Dev. Lett. **39**(1), 135–138 (2017)

[MNY12] S. Manipatruni, D.E. Nikonov, I.A. Young, Modeling and design of spintronic integrated circuits. IEEE Trans. Circuits Syst. I Regul. Pap. **59**(12), 2801–2814 (2012)

[Neu19] D. Neustadter. True random number generators for heightened security in any SoC (2019). https://www.synopsys.com/designware-ip/technical-bulletin/true-random-number-generator-security-2019q3.html

[Par+20] K.H. Park et al., High rate true random number generator using beta radiation, in *AIP Conference Proceedings*, vol. 2295. 1 (AIP Publishing LLC, 2020), p. 020020

[RPR17] N. Rangarajan, A. Parthasarathy, S. Rakheja, A spin-based true random number generator exploiting the stochastic precessional switching of nanomagnets. J. Appl. Phys. **121**(22), 223905 (2017)

[RS08] D.C. Ralph, M.D. Stiles, Spin transfer torques. J. Magn. Magn. Mater. **320**(7), 1190–1216 (2008)

[Ruk+01] A. Rukhin et al., NIST special publication 800-22. A statistical test suite for random and pseudorandom number generators for cryptographic applications (2001)

[Slo96] J.C. Slonczewski, Current-driven excitation of magnetic multilayers. J. Magn. Magn. Mater. **159**(1), L1–L7 (1996). ISSN: 0304-8853. https://doi.org/10.1016/0304-8853(96)00062-5. http://www.sciencedirect.com/science/article/pii/0304885396000625

[Sun06] J.Z. Sun, Spin angular momentum transfer in current-perpendicular nanomagnetic junctions. IBM J. Res. Dev. **50**(1), 81–100 (2006)

[SZ02] M.D. Stiles, A. Zangwill, Anatomy of spin-transfer torque. Phys. Rev. B **66**(1), 014407 (2002)

[UA+08] A. Uchida, K. Amano et al., Fast physical random bit generation with chaotic semiconductor lasers. Nat. Photonics **2**(12), 728–732 (2008)

[Vas+08] I. Vasyltsov et al., Fast digital TRNG based on metastable ring oscillator, in *International Workshop on Cryptographic Hardware and Embedded Systems* (Springer, Berlin, 2008), pp. 164–180

[Wei+16] Z. Wei et al., True random number generator using current difference based on a fractional stochastic model in 40-nm embedded ReRAM, in *2016 IEEE International Electron Devices Meeting (IEDM)* (IEEE, Piscataway, 2016), pp. 4–8

[XZS05] J. Xiao, A. Zangwill, M.D. Stiles, Macrospin models of spin transfer dynamics. Phys. Rev. B **72**(1), 014446 (2005)

[Yan16] C.A.O. Yang, Securing hardware random number generators against physical attacks. KU Leuven (2016). https://www.esat.kuleuven.be/cosic/publications/thesis-272.pdf

Chapter 6
Heterogeneous 2.5D and 3D Integration for Securing Hardware and Data

6.1 Chapter Introduction

2.5D and 3D integration involves vertically stacking and interconnecting multiple chips or active layers. This notion can be classified according to the underlying technology (Fig. 6.1):

(a) **Through-silicon via (TSV)-based 3D ICs:** Multiple chips are fabricated separately and then stacked and bonded, with vertical interconnects realized by TSVs, i.e., relatively large metal plugs that cut through the individual chips.

(b) **Face-to-face (F2F) 3D ICs:** Two chips are fabricated separately and then bonded directly at their BEOL metal "faces," with additional TSVs employed for global I/O as well as power and clock delivery.

(c) **Monolithic 3D (M3D) ICs:** Multiple active layers are manufactured sequentially, with vertical interconnects implemented by regular metal vias.

(d) **2.5D ICs:** Multiple chips are fabricated separately and integrated side-by-side on an integration and interconnects carrier, the interposer. These are also known as interposer stacks.

The two main factors driving 2.5D and 3D integration, in general, are (1) the CMOS scalability bottleneck, which becomes more exacerbated for advanced nodes due to issues like routability, pitch scaling, and process variations, and (2) the push toward advanced capabilities for heterogeneous and system-level integration, as required for more and more complex electronic systems. Both drivers are also known as "More Moore" and "More than Moore," respectively.

In fact, various studies, prototypes, and commercial products have shown that such 3D and 2.5D ICs offer significant benefits over conventional 2D ICs [Fic+13, Iye15, Kim+12, Shi18, PSF17, P D+17, Kim+19, Sto+17, Cle+16, Tak+13, Lau11, Viv+20]. The most important benefits offered by 3D and 2.5D integration are: (1) shorter, optimized system-level interconnects exploiting the third dimension, (2) reduced power and increased performance, bandwidth, (3) ease

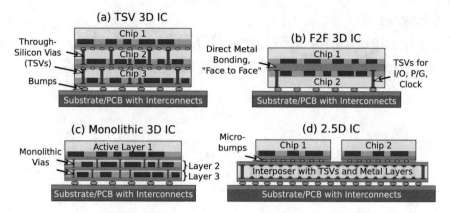

Fig. 6.1 Overview of 3D and 2.5D integration technologies

of heterogeneous, large-scale integration, along with the notion of IP reuse being raised to chip level, and (4) technology optimization for individual chips.

While 2.5D ICs do not offer the same integration density as native 3D ICs, building advanced electronic systems in the form of 2.5D ICs is, at this point in time, considered more promising [Sto+17, PSF17, Lau11]. This is also because interposers can be implemented using mature technology nodes, for improved cost savings and yield management. Concurrently, there are efforts for driving the notion of IP reuse toward the system level, also based on the 2.5D IC and interposer technology. Under these efforts, not only IP modules are to be reused at the chip level, but rather entire chiplets at the system level [Gre16, Yin+18, Sto+17, Kim+19, Lee+16, Int19]. Chiplets are relatively small chips encapsulating certain levels of complex functionality, like a microprocessor, as hard physical IP. The potential benefits of using chiplets are lower design and manufacturing costs, improved yield through separating technologies, and greater design flexibility. As such, the economic benefits, especially for small-volume development of heterogeneous and large-scale systems, are promising.

Concerning security, the notions of 2.5D and 3D integration cover both sides of the coin, i.e., promises as well as challenges for advancing security. In this chapter, we first discuss concepts for heterogeneous 2.5D and 3D integration that serve to advance hardware security, and we review related, selected prior art. Then, we present two case studies. The first study leverages the 3D IC technology and dedicated obfuscation of vertical interconnects for IP protection and prevention of targeted Trojan insertion, and the second highlights the 2.5D active interposer technology for secure system-level integration and operation of untrusted commodity chips. Note that throughout this chapter "foundry" and "fab" are used interchangeably, FEOL refers to front-end-of-line, and BEOL stands for back-end-of-line.

6.2 Concepts for 2.5D and 3D Integration for Security

The key concepts supported by heterogeneous 2.5D and 3D integration to advance security are as follows.

1. The relative **ease of separation** as well as **integration of trusted and untrusted components** (be it across interconnects, active devices, or both) serves to advance split manufacturing and integration of trustworthy monitoring circuitry with untrusted commodity hardware.
2. The **technology-agnostic nature** of vertical stacking enables heterogeneous integration, i.e., while mainstream 2.5D and 3D ICs are based on traditional CMOS technology, emerging devices can be utilized as well. This allows the designer to incorporate particular security solutions based on emerging devices, along with the security promises provided by 2.5D and 3D integration itself.
3. The relative ease of leveraging various **sources of entropy** in a 3D IC is particularly relevant for security. For example, advanced PUF architectures can be realized using the process variations exhibited within multiple chips in the stack, or using the physical phenomena incurred by vertical interconnects in general, or by leveraging coupling in particular [WYM14, Wan+15a]. Further, the power distribution networks of multiple chips experience different noise profiles, which can be exploited to mask power side-channel signatures within 3D ICs [DY18].
4. The ability to encapsulate sensitive components within larger, more complex stacks in 3D ICs enable the construction of **shielding structures**, rendering physical attacks more challenging [Bri+12a, KPS19a].

In the remainder of this chapter, we discuss in more detail how these concepts have been leveraged in prior art in general and selected case studies in particular.

6.3 Review of Selected Prior Art

As indicated, 2.5D and 3D integration offer various opportunities to advance hardware security. We provide a high-level overview on selected prior works in Table 6.1, and discuss more details in the following subsections.

6.3.1 Confidentiality and Integrity of Hardware

6.3.1.1 Camouflaging

The authors of [Yan+18] were the first to propose camouflaging dedicatedly for 3D integration, more specifically for M3D ICs. The authors developed and

Table 6.1 Selected works on 2.5D or 3D integration for hardware security

Reference	Style	Security scope; means	Trusted asset
[Val+13]	TSV	Runtime monitoring; Split manufacturing (SM)	Whole 3D IC
[XBS17]	2.5D	IP protection; SM	Passive interposer
[Ime+13b]	2.5D	Trojan mitigation; SM	Passive interposer
[Yan+18]	M3D	IP protection; Camouflaging	Whole 3D IC
[Pat+19b]	F2F	IP protection, Trojan mitigation; SM, camouflaging	Only BEOL
[Cio+14]	TSV	Probing protection; Physical enclosure	Whole 3D IC
[KS17], [BS19]	TSV	Side-channel mitigation; 3D integration by itself	Whole 3D IC
[Nab+20]	2.5D	Runtime monitoring; Separation of security backbone and commodity hardware	Active interposer

characterized custom M3D camouflaged libraries, and evaluated their scheme at the gate-level and chip-scale.

The physical embedding of camouflaging in [Yan+18] is realized by dummy contacts. Note that this has been proposed previously for camouflaging in 2D ICs. Particularly, the work in [Yan+18] leverages the benefits provided by M3D ICs in an effort to advance the scalability of camouflaging. That is noteworthy because prior art for camouflaging may incur considerable layout cost. For example, the NAND–NOR–XOR primitive of [Raj+13a] incurs $5.5\times$ power, $1.6\times$ delay, and $4\times$ area cost compared to a regular NAND gate. In practice, such overheads allow for only a few gates being camouflaged; in turn, the limited camouflaging scale renders such schemes prone to SAT-based attacks [Yu+17, Sha+17a]. In contrast, [Yan+18] reports, on average, only 25% power cost, 15% delay cost, and 43% area savings compared to regular 2D gates.

6.3.1.2 Split Manufacturing

Advancing split manufacturing via 3D and 2.5D integration seems both straight-forward and promising. This is because 3D and 2.5D integration allows to split a design into multiple chips, which can maintain their FEOL and BEOL layers independently as is, whereas the overall 2.5D/3D stack can comprise further parts of the system-level interconnects. Moreover, concerns regarding the practicality of conventional split manufacturing [Vai+14a, McC16] can be alleviated due to the fact that individual chips would not have to be split manufactured, but only the overall system.

The idea of such "3D split manufacturing" was first outlined in 2008, by Tezzaron Semiconductor [Tez08]. Various prior studies hinted at 3D split manufacturing as well, but most have some limitations. For example, [Dof+17] remained only on the conceptional level, while the studies [XBS17, Ime+13b] utilized 2.5D integration with "only" wires being hidden from untrusted facilities. The latter is in principal equivalent to traditional split manufacturing but is also more practical as it does not require to interrupt the regular manufacturing process. Nevertheless, the studies [XBS17, Ime+13b] reported considerable layout cost. Later works like [DRR17, Gu+18a, Pat+18c, Pat+19b] promoted "native 3D split manufacturing," i.e., with logic being split across trusted and untrusted facilities.

One important finding of those later studies [DRR17, Gu+18a, Pat+18c, Pat+19b] is that the 3D partitioning as well as the vertical interconnect fabric both play an important role in defining a cost-security trade-off as follows. The more the design is split up across multiple chips, the higher the layout cost (due to the need for more vertical interconnect links and related circuitry), but the more flexible and easier it becomes to "dissolve" the IP across the 3D stack. Note that [Gu+18a, Pat+18c, Pat+19b] proposed 3D split manufacturing in conjunction with camouflaging. While the study [Gu+18a] applied regular, FEOL-centric camouflaging, the studies [Pat+18c, Pat+19b] argued that another camouflaging approach is more appropriate for 3D split manufacturing, namely the obfuscation of the vertical interconnects (Fig. 6.2). For more details, see the related case study in Sect. 6.4.

Other works also suggest camouflaging at the level of system interconnects. For example, Dofe et al. [Dof+16] proposed to obfuscate the vertical interconnect fabric of 3D ICs by rerouting within dedicated network-on-chip (NoC) chips, "sandwiched" between the regular chips. This idea is conceptually similar to the notion of randomized routing in [Pat+18c, Pat+19b], but more flexible, yet also more costly.

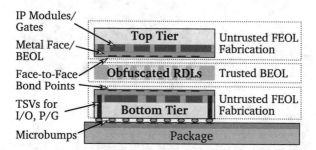

Fig. 6.2 Split manufacturing combined with camouflaging, as proposed in [Pat+18c, Pat+19b]. The splitting of the design across two tiers, along with obfuscation of the vertical interconnects, allows for IP protection and prevention of targeted Trojan insertion. This is because the two tiers reveal neither the design in its entirety nor individual components, to one or several colluding foundries

6.3.1.3 Hardware Trojan Defense

In [Pat+19b], the authors leveraged the benefits provided by 3D split manufacturing to advance the formally-secure but high-cost scheme of [Ime+13b] to mitigate Hardware Trojan (HT) insertion at manufacturing time. Besides such efforts, 3D and 2.5D ICs seem generally more vulnerable than 2D ICs to HT insertion, during design and manufacturing time. For example, the study in [MHE17] considered the negative bias temperature instability (NBTI) effect as a stealthy Trojan trigger, motivated by the fact that thermal management is a well-known challenge for 3D ICs.

In general, the broader landscape of suppliers and actors involved with 3D and 2.5D integration can open up new opportunities for attackers to embed HTs. Such a concern has also been voiced recently in [HCM19], along with the wide-spread adoption of wafer-level chip-scale packaging (WLCSP). Their attack hypothesized that some malicious integration facility could place a thin Trojan chip between the target chip and the package microbumps. The Trojan chip could contain TSVs to both pass-through and tap into those external connections, gaining access to all these signals at will (Fig. 6.3). To mitigate detection by visual or X-ray inspection, it was argued that aligning those TSVs with the microbump locations might suffice.

On the other hand, Trojan detection at runtime may benefit from 3D and 2.5D integration. This is because related security features can be implemented separately using a trusted fabrication process and integrated/stacked later on with the commodity chip(s) to be monitored [Wah+16, Nab+20]; see also the discussion on data security below.

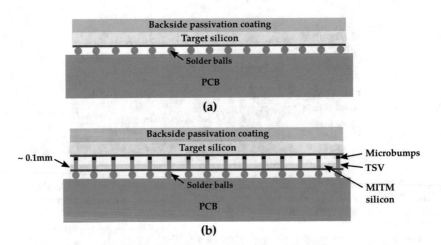

Fig. 6.3 Scenario for Trojan implantation through advanced packaging and 3D integration, as discussed in [HCM19]. The figure shows (**a**) a regular WLCSP chip, and (**b**) a man-in-the-middle (MITM) chip stacked between I/O bumps and the target chip, allowing an attacker to read all I/O data after deployment. Note that detecting such implants may be challenging in cost-optimized supply chain [HCM19]

6.3.1.4 Physically Unclonable Functions

As outlined before, 2.5D and 3D integration are particularly conducive for realizing PUFs. The readers are directed to Sect. 4.3.3 for further details on such 3D PUFs.

6.3.2 Data Security at Runtime

6.3.2.1 Unauthorized Access or Modification of Data

3D and 2.5D integration enables physical separation of components and, thus, allows for trustworthy realization of security features like runtime monitors [Mys+06, Val+13, Nab+20] or verifiers [Wah+16]. Still, the physical implementation of such schemes may become a vulnerability by itself. For example, in [Val+13], the authors propose introspective interfaces which require additional logic within the commodity chip to be monitored. It is easy to see that these interfaces would fail once they are modified by some malicious actor(s) involved with the design or manufacturing of that commodity chip. Thus, an undesirable dependency arises, possibly thwarting the scheme altogether. Note that the authors themselves acknowledge this limitation in [Val+13].

In [Nab+20], a 2.5D root of trust has been proposed, which integrates untrusted commodity chips/chiplets onto an active interposer which contains security features and further forms the backbone for system-level communication between the chips/ciplets. Thus, a clear physical separation into commodity and security components exists, avoiding any such security-undermining dependencies. For more details, see the related case study in Sect. 6.5.

6.3.2.2 Side-Channel Attacks and Physical Attacks

Side-channel attacks seem to become more difficult for 3D and 2.5D ICs due to the higher density of active devices and the increased complexity of circuit structures and architectures, which can result in more noisy side-channels. For example, the authors of [DY18] studied power side-channel attacks on 3D ICs, and observed that the power noise profiles from the different chips within the 3D IC are superposed. They also proposed a randomized cross-linking scheme of voltage supplies for cryptographic modules, to render attacks on such modules more difficult (Fig. 6.4).

Some studies leverage 3D and 2.5D integration to advocate for security schemes otherwise considered too costly. For example, the study in [BS15] leverages randomized eviction and heterogeneous latencies for a cache architecture, to prevent related side-channel attacks. The authors demonstrate that such techniques incur high-performance overheads in 2D ICs but can be realized gainfully in 3D ICs.

Other prior art studied side-channel attacks targeted explicitly at 3D ICs. An example is the thermal side-channel which is a somewhat limited but simple to

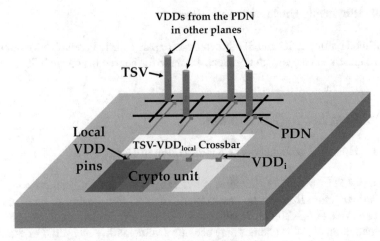

Fig. 6.4 Concept for mitigation of power side-channel leakage, based on a heterogeneous power-supply network driven by the multiple chips in a 3D IC. Derived from [DY18]

access proxy for the power side-channel [Mas+15]. For 3D ICs, the thermal side-channel can be mitigated by construction in different ways [Gu+16, KS17]; the heterogeneous nature of material composition and chip stacking, as well as the fact that chips are thermally cross-coupled are two important aspects toward that end. See Sect. 8.5 for more details and a related case study.

Fault-injection attacks, like side-channel attacks, may become more difficult due to the physical encapsulation of 3D/2.5D ICs. However, in [Rod+19] it was shown recently that a lateral re-arrangement of the laser setup may suffice to enable such fault-injection attacks against backside-protected 2D ICs and possibly also against 2.5D and 3D ICs.

The notion of physical enclosures enabled by 3D/2.5D integration may hinder read-out and probing attacks as well. In [KPS19a], the authors promoted an "all-around shield structure" enabled by 3D ICs, more specifically through TSVs placed densely at the chip boundaries, along with regular shields in the BEOL and backside protection. Similar protection against probing has been envisioned before in [Bri+12a, Cio+14]. While powerful in principle, such schemes are yet to be demonstrated in practice.

6.4   Case Study I: 3D Integration for IP Protection and Trojan Mitigation

Split manufacturing and layout camouflaging are two promising techniques to obscure ICs from malicious entities during and after manufacturing (also see the background for these techniques in Sects. 1.1.2.2 and 1.1.2.3). While both

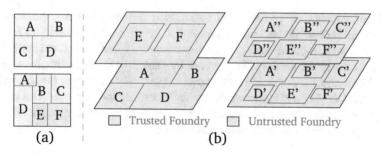

Fig. 6.5 (a) Current chip version (top) versus new chip version (bottom). For the new version, the IP modules *E* and *F* are entirely new, while the other modules are revised and/or reshaped. (b) Foundry scenarios for IP protection scheme using 3D ICs. For both tiers manufactured by an untrusted foundry (right), IP modules can be split up for obfuscation

techniques enable protecting the IP of ICs, they provide protection against different adversaries; combining the benefits of both seems interesting and relevant, especially when considering the following scenario. Lacking its own fabrication facilities, a company commissions a potentially malicious foundry to manufacture their latest chip version. This new version is typically extended from some previous version (Fig. 6.5a)—the reuse of IP modules and the re-purposing of proven architectures are well-known practices. For example, think of the flagship iPhone® by Apple®. The iPhone 7, based on the A10 chip, was launched in September 2016, and the iPhone X, based on the successor chip A11, was launched in September 2017. In such a scenario, it is intuitive that pirating the new IP can become significantly less challenging for fab-based adversaries. In case the same fab was already commissioned for the previous chip version, they readily hold the layout of that earlier version; otherwise, the adversaries can apply RE on chips of the previous version bought in the market. In any case, the adversaries can compare that new layout with the prior layout, to locate and focus on those parts which are different and unique.

The conclusion from this example is that both layout camouflaging and split manufacturing are required for manufacturing all the different chip versions. Layout camouflaging is required to prevent RE of the current layout by any other fab commissioned for later chip versions, whereas split manufacturing is necessary to prevent the fab which is manufacturing the current version (and which is also tasked to implement layout camouflaging) from readily inferring the complete layout of the current version. Prior art can only account for this scenario by applying split manufacturing on top of layout camouflaging, which can exacerbate the individual overheads and shortcomings, as discussed in Sect. 6.4.3 in more detail.

Furthermore, split manufacturing can also mitigate the insertion of HTs (also see the background on HTs in Sects. 1.1.2.3 and 1.1.2.4), especially in the context of 3D integration. That is because, for one, components considered prone to HT insertion can be delegated to trusted facilities for fabrication of separate chips, hindering HT insertion to begin with. For another, when an a priori decision on which components are prone to HT insertion is difficult to make, the notion of obfuscating vertical interconnects (covered in detail in this case study) helps to implement other, foundationally secure schemes with superior cost and scalability.

In this case study, we describe the work by Patnaik et al. [Pat+19b, Pat+18c] in detail. The authors argue that 3D integration serves well to combine the strengths of layout camouflaging and split manufacturing in one scheme (Fig. 6.2). The key idea is to "3D split" the design into multiple tiers and to obfuscate (i.e., randomize and camouflage) the vertical interconnects between those tiers.

6.4.1 3D Integration and Implications for Security and Practicality

The first step in leveraging 3D integration to advance selected security schemes is to study the integration options and the related implications for security and practicality. Thereafter, in the subsequent subsections, this case study covers the details on implementation and evaluation for these two security schemes of choice, i.e., IP protection and HT mitigation.

The primary objective which Patnaik et al. propose for split manufacturing is to "3D split" the design into multiple tiers. That is, unlike regular split manufacturing in 2D where the layout is split into FEOL and BEOL, the authors split the design itself into two parts (or more, in principle). These parts are manufactured as separate chips, and then stacked and linked by obfuscated vertical interconnects following the F2F integration process; this technology choice is without loss of generality but appropriate as discussed later on in more detail. Patnaik et al. suggest that such 3D split manufacturing can be achieved either by commissioning different foundries or one foundry (Fig. 6.5b), in the following scenarios.

1. **Different trusted and untrusted foundries:** Consider one trusted and one untrusted foundry, both with full FEOL and BEOL capabilities, but for different technology nodes. It is intuitive to delegate the sensitive parts to the trusted fab exclusively. While this approach is straightforward and inherently secure against fab-based adversaries, its practicality is limited.
2. **Untrusted foundries/foundry:** Consider one or more high-end but untrusted fab(s). This way, one can benefit from the latest technology but, naturally, has to obfuscate the design in such a way that the fab(s) cannot readily infer the whole layout, even when they are colluding. Once such strong protection is in place, it is economically more reasonable to commission only one fab.

Next, both scenarios are elaborated on in the remainder of this section. However, for the case study itself, the focus is on the more relevant and practical scenario 2. Besides, to further achieve security against (a) fab-based adversaries and (b) malicious end-users, the case study looks into (a) randomizing the vertical interconnects and (b) obfuscating those interconnects. Therefore, only a trusted BEOL is required, but no trusted FEOL facility.

6.4.1.1 Combining Different Trusted and Untrusted Foundries

Commissioning several foundries, providing different trust levels, and supporting different technology nodes hold two key implications as follows.

First, it is intuitive to assign sensitive design parts to the chip manufactured by the trusted foundry. Examples of sensitive parts can include (1) some new IP to protect (Fig. 6.5b), or (2) parts considered vulnerable to targeted HT insertion (Fig. 6.6a). Such an approach is secure by construction against fab-based adversaries for the following reasons. For IP piracy, there is no generic attack model in the literature yet which can infer missing connections and gates, when given only a part of the overall design. One may argue that such "black-box attacks" would be very challenging, if possible at all. For HT insertion, the adversary cannot perform targeted insertion once the vulnerable parts are delegated to the trusted chip exclusively.

Second, in case a trusted fab and another untrusted fab are commissioned in parallel, it is implied that these two fabs would support different nodes, with the trusted fab typically offering access only to an older technology. In fact, if the trusted foundry would be able to offer the same high-end node, one could simply commission the trusted foundry for manufacturing of the whole design. Due to the different pitches for different technology nodes, however, only a fraction of the design can be delegated to the trusted low-end fabrication. Also, power and performance will be dominated by the low-end chip, where factors such as parasitics, level shifting, and clock synchronization may further exacerbate the overheads [Pen+17, GM13].

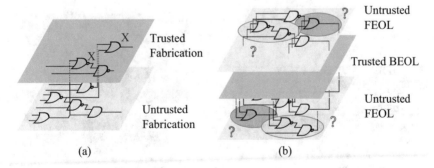

Fig. 6.6 Two approaches to using 3D integration for the prevention of targeted HT insertion. (**a**) Assuming a trusted foundry (along with another untrusted one), the vulnerable design parts (marked by "X") can be delegated to the trusted foundry. (**b**) Assuming only an untrusted FEOL facility, the design has to be split in such a way that an FEOL-based attacker cannot readily identify the vulnerable parts. This also requires randomization of the vertical interconnects using a trusted BEOL facility

Experimental Evaluation on MIT-LL Common Evaluation Platform Patnaik et al. study the scope for such heterogeneous 3D split manufacturing as follows. The authors study a scenario where the sensitive logic is moved to a trusted fab, offering the older 90 nm technology, whereas the remaining logic is delegated to an untrusted 45 nm fab. This study is based on multi-million-gate System-on-Chip (SoC), provided by MIT-LL as a Common Evaluation Platform (CEP) [CEP19]. An overview of the SoC architecture is given in Fig. 6.7a; it is a one-master ten-slaves system. The master is a re-engineered version of the *OpenRISC* processor, called OR1200. This OR1200 master executes code from a 128KB static RAM (SRAM), which is considered to be available off-chip. The slaves comprise cryptographic (crypto) modules, digital signal processing (DSP) modules, and a global positioning system (GPS) processing module. The Wishbone/DBUS connects the processor master with all other blocks, while UART provides a serial interface for off-chip communication.

Given that all considered crypto modules are public knowledge, the designer (attacker) would not be interested in protecting (retrieving) the related IP. However, the DSP modules and other parts may contain customized logic and sensitive IP worth protecting. As discussed before, the choice of which modules/logic to protect lies solely with the design house, as they can best judge which components require IP protection. For their case study, Patnaik et al. assume the GPS processing module to be that sensitive asset.

Patnaik et al. use the *NanGate* 45 nm library [NG11] and *Synopsys* 90 nm library [Synopsys32nm19]. *Cadence Innovus 17.1* is used for layout generation and PPA evaluation; the setup details are further elaborated in Sect. 6.4.3.2. First, the authors synthesize the 2D baseline designs, considering the slow corners for the 45 and 90 nm nodes, respectively; the results are given in Table 6.2. Second, the authors partition the 45 nm baseline design such that all logic of the GPS module

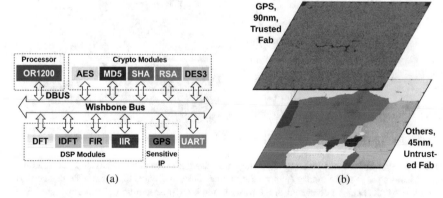

(a)   (b)

Fig. 6.7 (a) Architecture of the MIT-LL Common Evaluation Platform [CEP19]. The GPS block is considered the sensitive IP; it is thus to be delegated to the trusted fabrication. (b) Heterogeneous 3D implementation, with colors corresponding to modules in (a)

Table 6.2 PPA numbers for separate 2D baselines, using the *NanGate* 45 nm library [NG11] and the *Synopsys* 90 nm library [Synopsys32nm19]. Layouts are DRC clean, and have a utilization of 70%. Area is in μm², power in mW, and delay in ns

Benchmark	45 nm				90 nm			
	Inst.	Area	Power	Delay	Inst.	Area	Power	Delay
CEP	842,685	1,568,893	429.91	4.41	801,814	10,582,015	1488	5.84

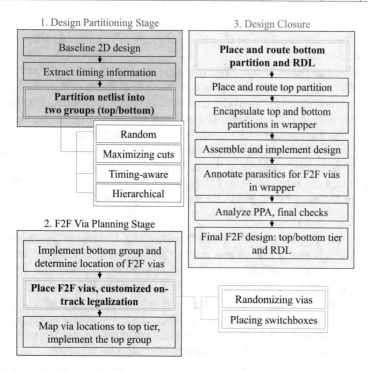

Fig. 6.8 CAD flow for F2F 3D ICs, proposed by Patnaik et al., implemented in *Cadence Innovus*. Security-driven steps for IP protection are emphasized in bold

can be delegated to a trusted 90 nm chip. The authors follow the design flow shown in Fig. 6.8 in general, but there are some differences as follows: (a) they have to re-synthesize the GPS module for 90 nm, with the same timing constraint as for the remaining modules, which is also to simplify the clock-tree synthesis (CTS) for the individual tiers [GM13]; (b) for layout evaluation, they revise the LIB and LEF files for the GPS tier; (c) they do not undertake any additional security-centric steps (highlighted in bold in Fig. 6.8) as the GPS IP is fully secured against adversaries residing in the 45 nm fab. Finally, they assume supply voltages of 0.95 V for 45 nm and 0.9 V for 90 nm when generating the power numbers. Note that, in case different parts require considerably more different voltages, level shifters should be used. Along with technology scaling, however, the nominal voltage has stagnated (since

Table 6.3 PPA numbers for the heterogeneous 3D implementation of Fig. 6.7b, with the GPS module being assigned exclusively to the 90 nm tier. Layouts are DRC clean, and have a utilization of 70%. Area is in μm^2, power in mW, and delay in ns

Benchmark	Area				Power	Delay
	45 nm Inst.	45 nm area	90 nm Inst.	90 nm area		
CEP	708,858	1,409,486.12	131,342	1,149,300.63	560.81	5.08

90 nm), which allows most modern processes to run at compatible voltages without shifters [Gu+18b].

The PPA numbers for this 3D implementation (with different-foundries) are given in Table 6.3. One can observe overheads of 63.09%, 30.45%, and 15.03% for area, power, and delay, respectively, compared to the 2D 45 nm baseline. However, when comparing the 3D implementation with the trusted 90 nm baseline, their approach offers savings of 75.82%, 62.31%, and 13.1% for area, power, and delay, respectively. Also note that the 90 nm GPS tier comprises only 15.63% of the total number of instances, although it incurs almost the same area footprint. In short, heterogeneous 3D integration may indeed provide benefits over a purely trusted fabrication (90 nm), but it naturally cannot compete with advanced fabrication (45 nm).

Summary for Scenario of Different Trusted and Untrusted Foundries This approach of utilizing a trusted and another untrusted foundry for 3D integration may be inherently secure against fab-based adversaries, and may also offer some PPA benefits over a 2D implementation using only the trusted and old node, but it is also limited in practice. That is because one can delegate only a small fraction of the overall design to the trusted foundry (i.e., at least without incurring area cost), which limits the scale for IP protection, and the performance and power of this 3D approach cannot compete with the advanced but untrusted 2D node.

6.4.1.2   Using Untrusted Foundries Solely

Engaging with (1) several untrusted foundries offering the same technology node or (2) one untrusted foundry, also holds some key implications as follows.

First, power and performance of such "conventional" 3D integration can be expected to excel those of the heterogeneous scenario above. Splitting of 2D IP modules within 3D ICs has been successfully demonstrated, e.g., in [Jun+17], albeit without hardware security in mind. Hence, savings from the folding of IP modules may provide some margin for security schemes. However, it is also shown in detail that this margin naturally depends on the design and the measures applied for the scheme.

Second, although IP modules can be split across tiers, which may mislead an RE attacker (a malicious end-user), both tiers are still manufactured by some untrusted fab(s). This fact implies that layout camouflaging schemes targeting the device level cannot protect from adversaries in those foundries. Interestingly, there is another flavor of layout camouflaging discussed recently that is the obfuscation of interconnects [Che+15, Pat+17]. Patnaik et al. argue that obfuscation of interconnects is a natural match for F2F 3D integration—in between the two tiers, redistribution layers (RDLs) can be purposefully manufactured for obfuscation of the vertical interconnects (Fig. 6.2). Doing so only requires a trustworthy BEOL facility, which is a practical assumption given that BEOL fabrication is much less demanding than FEOL fabrication (owing to larger pitches and less complex processing steps). This is especially true for higher metal layers; note that RDLs reside between the F2F bonds which themselves are at higher layers.

Chen et al. [Che+15] consider real and dummy vias using Magnesium (Mg) and Magnesium Oxide (MgO), respectively, for obfuscation of interconnects. They demonstrate that real Mg vias oxidize quickly into MgO and, hence, can become indistinguishable from the other MgO dummy vias during RE. Without loss of generality, Patnaik et al. assume their layout camouflaging scheme to be based on the use of Mg/MgO vias for obfuscating the vertical interconnects of the 3D IC. Emerging interconnects such as those based on carbon nanotubes [Uhl+18] may become relevant in the future as well.

6.4.2 Methodology for IP Protection

Here, the CAD and manufacturing flow for F2F 3D integration, as proposed by Patnaik et al., is discussed. The CAD flow is in parts inspired by Chang et al. [Cha+16], but independently customized and implemented, with a particular focus on IP protection (Fig. 6.8). The flow allows a concerned designer to explore the trade-offs between PPA and cuts, i.e., the number of F2F vertical inter-tier connections. Cuts are a crucial metric for the security analysis, as discussed in more detail in Sect. 6.4.3.4. Note that the authors purposefully do not engage cross-tier optimization steps, to mitigate layout-level hints on the obfuscated BEOL/RDLs.

As for the F2F process, Patnaik et al. propose the following security-centric modification. The wafers for the two tiers are fabricated by one (two) untrusted foundry (foundries) and then shipped to a trusted BEOL and stacking facility. This trusted facility grows the obfuscated RDLs on top of one wafer and continues with the regular F2F flow (i.e., flipping and bonding the second wafer on top).

6.4.2.1 Design Partitioning

After obtaining the post-routed 2D design, the netlist is partitioned into top and bottom groups, representing the tiers of the F2F IC. I/O ports are created for all vertical interconnects between the two groups, representing the F2F vias. Besides these F2F ports, primary I/Os are placed at the chip boundary, as in conventional

2D designs. This is also practical for F2F integration where TSVs are to be manufactured at the chip boundary for primary I/Os and the P/G grid.

Random Partitioning A naive way for security-driven partitioning is to assign gates to the top/bottom groups randomly. While doing so, the number of cuts will be dictated by the number, type, and local inter-connectivity of gates being assigned to one group. Since random partitioning lacks any heuristic, it may either result in savings or overheads for power and/or performance, depending on the design, number of vertical interconnects induced, and randomness itself.

Maximizing the Cut Size As already indicated (and further explored in Sect. 6.4.3.4), the larger the cut size, the more difficult becomes IP piracy. Hence, Patnaik et al. seek to increase the cut size as much as reasonably possible. First, timing reports for the 2D baseline are obtained following which gates are randomly alternated along their timing paths toward the top/bottom groups. In the security-wise best case, which is also the worst case regarding power and performance, every other gate is assigned to the top and bottom group, respectively. There, for a path with n gates, $2n$ cuts are arising. In short, the trade-off is as follows: The larger the cut size, the more resilient the design, but the higher the layout cost. The impact of maximizing the cut size is discussed in Sect. 6.4.3.2.

Timing-Aware Partitioning Based on the insights regarding the cost-security trade-off for random partitioning and maximizing cuts, Patnaik et al. seek to reduce layout cost while maintaining strong protection. First, the available timing slack is determined for each gate. Then, based on a user-defined threshold, the critical gates remain in the bottom tier, whereas all other gates are moved to the top tier. This procedure is repeated with revised timing thresholds until an even utilization for both tiers is achieved. Note that it is difficult for an attacker to understand whether a path in the bottom/top group is critical or not (or even complete, for that matter). In other words, the attacker has to tackle both groups at once and, more importantly, resolve the randomized F2F vias and the obfuscated interconnects.

Patnaik et al. advocate this partitioning strategy, especially for any flat design. In the remainder of this case study, timing-aware partitioning is the default strategy considered, unless otherwise noted.

Hierarchical Partitioning This strategy is applied for designs with hierarchies in the top-level module. Inspired by Chang et al. [Cha+16], Patnaik et al. separate modules with a large degree of connectivity across tiers, resulting in large numbers of cuts. Other modules are partitioned/placed to balance the utilization of both tiers. In short, this strategy serves to protect the IP as well as to limit layout cost for hierarchical designs.

6.4.2.2   Planning of F2F Interconnects

After placing the bottom tier, the initial locations for F2F ports are determined in the vicinity of the drivers/sinks. Then, a security-driven, randomized placement of F2F ports is conducted, along with customized on-track legalization. Next, obfuscated switchboxes are placed, and the F2F ports are mapped to the top tier.

Randomization It is easy to see that regular planning of F2F interconnects cannot be secure, as this aligns the ports for the bottom and top tier directly. That is, the untrusted foundry has direct access to both tiers and could simply stack them up to recover the complete design. Hence, the arrangement of F2F ports is randomized as follows (see Fig. 6.9). They place additional F2F ports randomly, yet with help of on-track legalization (see below), in the top RDL. These randomized ports are then routed through the RDLs toward the original F2F ports connecting with the bottom tier. In short, randomization of F2F vias is required to protect the design against fab-based adversaries during manufacturing.

Obfuscated Switchboxes To further protect against RE attacks from malicious end-users, Patnaik et al. obfuscate the connectivity in the RDLs, using a customized switchbox structure (Fig. 6.10). This switchbox allows stealthy one-to-one mapping of four drivers to four sinks. The essence of the switchbox is the use of Mg/MgO vias (recall Sect. 6.4.1.2), to cloak which driver connects to which sink. The pins of the switchbox represent the F2F ports. The pins are aligned with the routing tracks to enable proper utilization of routing resources. For randomization, the additional ports connecting with the top tier are used for rerouting during design closure.

On-Track Legalization Each F2F port is moved inside the core boundary, toward the center point defined by all instances connected with this port. Next, the closest and still-unoccupied on-track location for actual placement is obtained. If need be, stepwise increase of the search radius considering a user-defined threshold is conducted.

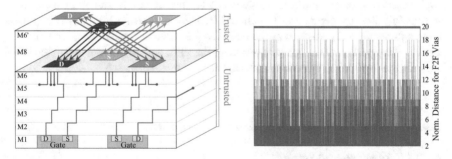

Fig. 6.9 (Left) RDL randomization for switchboxes and F2F vias. (Right) Normalized distances between to-be-connected F2F vias after randomization, for benchmark *b17_1*

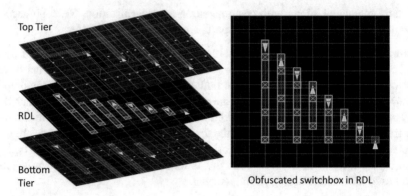

Fig. 6.10 Obfuscated switchbox, embedded in two RDL layers, exemplarily for bottom-to-top drivers. Each driver pin (downwards triangle) can connect to any sink pin (upwards triangle). For simplicity, all F2F ports are aligned here with the pins of the switchbox, whereas the top-tier ports are randomized in reality

6.4.2.3   Design Closure

After the F2F via planning stage, both tiers are placed and routed separately, independent of each other. For sequential designs, Patnaik et al. conduct CTS on both tiers independently, as suggested in [GM13].

Also recall that the authors do not engage in any cross-tier optimization, on purpose, to anonymize the individual tiers from each other. However, they apply intra-tier optimization. While routing the bottom tier, they also route the randomized and obfuscated RDL with their switchboxes. Next, the authors encapsulate the top and bottom partitions in a wrapper netlist, and assemble and implement the design followed by generating a Standard Parasitic Exchange Format (SPEF) file that captures the RC parasitics of the F2F vias (modeled as regular vias, see below). Finally, they perform DRC checks, evaluate the PPA, and stream out separate DEF files for the top/bottom tiers and the RDL.

6.4.3   Results and Insights for IP Protection

6.4.3.1   Experimental Setup

Implementation and Layout Evaluation   Since there are no commercial tools available yet for (F2F) 3D ICs, Patnaik et al. implement their CAD flow within *Cadence Innovus 17.1*, using custom *TCL* and *Python* scripts. Their implementation imposes negligible design runtime overheads. The authors use the *NanGate* 45 nm library [NG11] for their experiments, with six metal layers for the baseline 2D setup and six layers each for the top and bottom tier in the F2F setup. The RDL comprises four duplicated layers of M8, from which two are used for embedding the obfuscated switchboxes, and two are used for randomizing the routing. F2F vias are modeled as M6 vias; while this is an optimistic assumption, for now, F2F technology scaling can be expected to reach such dimensions. The PPA analysis is conducted for the slow process corner, using *CCS* libraries at 0.95 V. For power analysis, the authors assume a switching activity of 0.2 for all primary inputs. The authors ensure that the layouts are free of any congestion, by choosing appropriate utilization rates for the 2D baselines. This is essential to prevent any possible congestion to be carried forward in their 3D flow. All experiments are carried out on an Intel Xeon E5-4660 @ 2.2 GHz with *CentOS 6.9*. For *Cadence Innovus*, up to 16 cores are allocated.

Setup for Security Evaluation Since Patnaik et al. promote 3D split manufacturing, regular proximity attacks such as [RSK13, Wan+18b] cannot be applied. Thus, the authors propose and publicly release a novel attack against 3D split manufacturing [Kne18], also accounting for the RDL obfuscation underlying in their scheme; see also Sect. 6.4.3.4. Attacks on their protected layouts are evaluated by commonly used metrics, i.e., the correct connection rate (CCR), percentage of netlist recovery (PNR) [Pat+18e], and Hamming distance (HD). HD is calculated using *Synopsys VCS* with 1,000,000 test patterns. As for SAT-based RE attacks, the authors leverage the tool provided by [SRM15], with the related time-out set to 72 h.

Designs The commonly considered benchmarks from the *ISCAS-85* and *ITC-99* suites are used for layout and security analysis. In addition, two SoC benchmarks are used: the MIT-LL CEP [CEP19] and the JPEG OpenCores design [OC19].

6.4.3.2   Security-Driven Layout Evaluation

The flow allows to trade-off PPA and cuts; the latter dictates the resilience against IP piracy both during and after manufacturing. Figure 6.11 showcases the layout images for benchmark *b22*.

Random Partitioning and Maximizing the Cut Size Initially, Patnaik et al. study random partitioning of gates, by moving them randomly from the bottom to the top group in steps of 10%, up to 50%. As the strategy is randomized, the authors perform ten runs for each benchmark for any given percentage of gates to move. The resulting power and performance distributions are illustrated in Fig. 6.12.

Interestingly, even for the security-wise best case of randomly moving 50% of the gates, some runs still provide better power and/or performance than the 2D baseline. The savings in performance can be attributed to the fact that, when splitting the design across the vertical dimension, one can obtain a reduction in wirelength, which helps to improve timing. Note that these improvements, on average, come at some expense of power, with related overheads in the range of 0–7% for lifting/moving 50% of gates.

While this demonstrates the potential for naive random partitioning, it is important to note that this finding only holds true as long as one refrains from randomizing

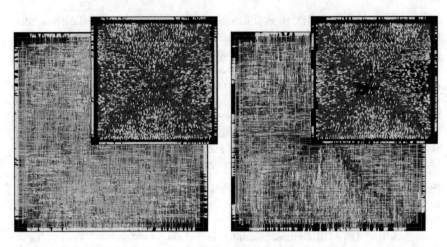

Fig. 6.11 Layout snapshots of bottom/top tier (left/right) for *b22*. The insets show the corresponding F2F vias

the F2F ports and from using the obfuscated switchboxes for these experiments. In fact, once the authors seek to maximize the cuts, along with randomization of F2F ports and use of switchboxes, larger *ITC-99* benchmarks such as *b18_1* incur considerable overheads of up to 60% (Fig. 6.13). Here Patnaik et al. also observe that large cut sizes lead to an increase in routing congestion and total wirelength, thereby further increasing the total capacitance of the design. This offsets the performance benefits which regular, security-oblivious 3D integration can be expected to achieve [KCL18].

In short, although these strategies offer strong resilience, a more aggressive PPA-security trade-off may be desired.

Timing-Aware Partitioning This setup tackles that need for achieving security while maintaining reasonable PPA cost. Patnaik et al. observe that even for larger *ITC-99* benchmarks such as *b18_1* and *b19* (Fig. 6.14), there are some benefits when comparing the secure 3D designs to their 2D baseline. As explained in Sect. 6.4.2.1, since the most timing-critical gates are constrained to one tier, the authors induce significantly less cuts along the timing paths for the 3D design. For example, the authors observe a reduction of about 60% in timing-path cuts for *ITC-99* benchmarks *b18_1* and *b19* when compared to random partitioning. To demonstrate the security implication of this setup, the authors plot the normalized distances between to-be-connected F2F vias in Fig. 6.9. This figure shows a wide variation across the inter-tier nets, whereas for regular, unprotected F2F stacking the distances would be all zero. Overall, the choice of partitioning lies with the designer, which she/he can trade-off considering security and PPA cost, but timing-aware partitioning should be considered first, i.e., at least for non-hierarchical designs.

Experimental Evaluation on CEP and JPEG Besides the well-known benchmarks considered above, the authors also considered two "real-world netlists," the

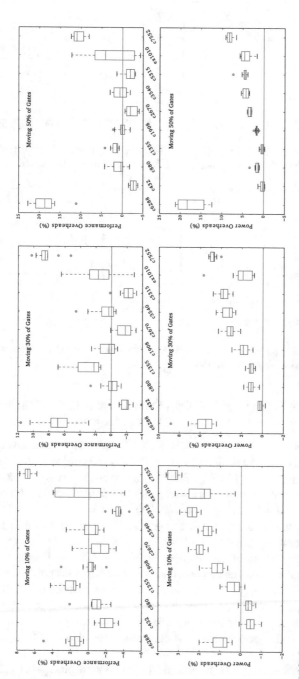

Fig. 6.12 Impact of randomly assigning gates on performance (top) and power (bottom). Each boxplot represents ten runs. Note that the same benchmarks are applied for the top and bottom plots; benchmark labels are accordingly placed between those plots

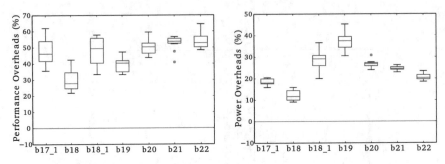

Fig. 6.13 Layout cost for maximizing cuts, with 35–50% of the gates moved, and with obfuscated switchboxes and F2F randomization being applied. Each boxplot represents ten runs

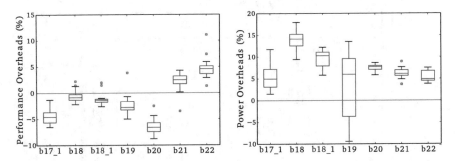

Fig. 6.14 Performance, power cost for timing-aware setup with obfuscated switchboxes, and F2F randomization. Each box represents ten runs

MIT-LL CEP [CEP19] and the JPEG OpenCores design [OC19]. Thus, the authors also demonstrate their secure end-to-end CAD flow for practical 3D ICs.

The utilization is set to 70% and 60% for CEP and JPEG, respectively, which ensures that the 2D baseline designs are devoid of any congestion. Patnaik et al. use the *NanGate* 45 nm library [NG11]. F2F vias are modeled as M10 vias. All 2D and 3D designs operate at iso-performance, with a timing constraint of 5 ns (i.e., at 200 MHz). Further details are the same as in Sect. 6.4.3.1.

The results for the 2D baseline and secure 3D designs are provided in Table 6.4. Regarding the footprint area/die outlines, both the secure 3D designs provide savings over their 2D baselines, namely 49.3% for CEP and 42% for JPEG. Regarding instance counts, one can observe some overheads for both the secure 3D designs; as they do not apply any cross-tier and/or post-partitioning optimization, there is less leverage to reduce instance counts for the tools. Again, for regular, security-oblivious 3D F2F integration, one would expect savings/reductions in both wirelength and instance count, which ultimately also enables power savings [KCL18].

For their security-driven 3D flow, it depends on various aspects whether there are power/performance savings or overheads. First, recall that the authors randomize

Table 6.4 Comparison between 2D baseline designs and their secure 3D F2F counterparts of CEP [CEP19] and JPEG [OC19]. All results are for iso-performance (5 ns). Wirelengths in 3D are subject to randomization of F2F vias

Metrics	CEP		JPEG	
	2D	3D	2D	3D
Footprint (μm^2)	2,332,429	1,180,807	1,317,041	763,945
F2F via count	0	6,447	0	6,707
Total cell count	852,496	859,207	497,666	520,374
Total wirelength (m)	26.62	23.08	8.69	9.58
Total power (mW)	417.8	390.7	204.1	221.9

F2F vias and leverage obfuscated switchboxes (to deter fab-based adversaries and malicious end-users, respectively), which tends to increases wirelength, and thereby the driver strengths and/or buffer counts. Second, the designs have some impact by themselves. For example, for JPEG the authors note 8.72% higher power consumption, whereas for CEP they note a 6.49% power reduction. Third, partitioning plays an important role as well, as already discussed. Since CEP and JPEG are both hierarchical designs, Patnaik et al. apply hierarchical partitioning, which fully protects the system-level IP orchestration and any glue logic. To further protect individual modules, one can split them up across the two tiers: which modules to select and how to split is the designer's decision, also depending on the nature of the modules and the overall design [Jun+17]. Toward this end, the authors also performed an experiment on CEP where individual modules were partitioned, resulting in 20,863 F2F vias (3.25× than those reported in Table 6.4). Patnaik et al. maintain that both tiers are DRC-clean and free of congestion. The authors observe power and timing overheads of 9.59% and 13.64%, respectively, which implies that this 3D design can operate at around 176 MHz. Running the overall chip at this frequency also ensures that there is no loss in system functionality. Finally, note that the number of cuts obtained here indicate a strong resilience of the 3D designs; this is discussed further in Sect. 6.4.3.4.

6.4.3.3   Comparison with Prior Art

Layout Camouflaging Schemes Among others, threshold-voltage-dependent layout camouflaging is gaining traction. Although promising concerning resilience, the PPA cost are considerable. For example, Akkaya et al. [AEM18] report overheads of 9.2×, 6.6×, and 3.3× for PPA, respectively, when compared to conventional 2-input NAND gate. Nirmala et al. [Nir+16] report 11.2× and 10.5× cost for power and area, respectively. Besides, for interconnects camouflaging, Patnaik et al. [Pat+17] report PPA overheads of 4.9%, 31.2%, and 25% for *ITC-99* benchmark *b17* at 60% layout camouflaging. When compared to these schemes, Patnaik et al. can provide significantly better PPA (except for [Pat+17] concerning power).

Regarding prior art on 3D layout camouflaging [Yan+18, Gu+18b], recall that they require a trusted FEOL facility; hence, their schemes are not directly comparable to that of Patnaik et al. Also, at the time of compiling this case study, their libraries and protected designs were not available to the authors for a detailed study. Moreover, for [Gu+18b], the authors leverage regular 2D layout camouflaging schemes while using different technology nodes. Depending on the particular node and layout camouflaging scheme, this may induce large PPA overheads, and technology-heterogeneous 3D integration may hold further complications [Pen+17, GM13]. Recall that these concerns were their main motivation to advocate the use of uniform/same technologies and camouflaging of vertical interconnects.

Split Manufacturing Schemes In Table 6.5, Patnaik et al. compare with studies on 2D split manufacturing. Overall, the placement-centric techniques by Wang et al. [Wan+18b] are competitive concerning power and performance. However, as always the case for regular split manufacturing, they can only avert fab-based adversaries, but not malicious end-users.

6.4.3.4   Security Analysis and Attacks

Proximity Attack for 3D Split Manufacturing Patnaik et al. propose and implement an attack, which can account for 3D split manufacturing in the context of IP piracy, with a focus on one untrusted foundry (or two colluding foundries) and their RDL obfuscation. They provide this attack as a public release in [Kne18].

Patnaik et al. assume that the attacker holds the layout files for the top and bottom tier, but, residing in the untrusted fab, she/he has no access to the trusted RDL. Note that the implications for malicious end-users being able to access the obfuscated RDL are discussed further below. Although she/he understands how many drivers are connecting from the bottom to the top tier and vice versa, she/he does not know which driver connects to which sink, given the randomization of F2F vias. Recall that, the authors do not engage in cross-tier optimization, to mitigate any layout-level hints. Assume there are d_{bot} drivers in the bottom and, independently, d_{top} drivers in the top tier. Since the authors do not allow for fan-outs within the RDL (this would occupy more F2F vias than necessary), there are only one-to-one mappings—this results in $d_{bot}! \times d_{top}!$ possible netlists. Once switchboxes are used, however, the attacker can tackle groups of four drivers/sinks at once. Still, she/he has to resolve (a) which four top-tier drivers are connected to which four bottom-tier sinks and vice versa, and (b) the connectivity within the obfuscated switchboxes. For those cases, there are $4! \times \left((1/4 \times d_{bot})! \times (1/4 \times d_{top})!\right)$ possible netlists remaining. Next, the corresponding heuristics for the attack are outlined.

1. *Unique mappings:* Any driver in the bottom/top tier will feed only one sink in the top/bottom tier. Hence, an attacker will reconnect drivers and sinks individually.

Table 6.5 PPA cost comparison with 2D split manufacturing protection schemes. Numbers are in % and quoted from the respective publications

Benchmark	BEOL+Physical [Wan+18b]			Logic+Physical [Wan+18b]			Logic+Logic [Wan+18b]			Concerted lifting [Pat+18e]			Proposed with random partitioning		
	Area	Power	Delay	Area	Power	Delay	Area	Power	Delay	Area	Power	Delay	Area[a]	Power	Delay
c432	N/A	0.17	0.49	N/A	0.44	0.24	N/A	0.17	0.21	7.7	13.1	11.6	−50	−2.66	0.31
c880	N/A	0.25	0.05	N/A	0.35	0.03	N/A	−0.05	−0.09	0	12.1	19.9	−50	0.97	1.6
c1355	N/A	0.52	0.57	N/A	0.75	0.42	N/A	0.03	0.01	0	12.2	21.3	−50	1.83	0.38
c1908	N/A	1.1	1.3	N/A	1.1	0.23	N/A	0.45	0.39	7.7	14.6	18.9	−50	0.11	1.69
c2670	N/A	0.29	0.27	N/A	0.29	0.27	N/A	0.05	0.03	7.7	10	12	−50	−2.18	3.32
c3540	N/A	0.53	0.28	N/A	0.36	0.02	N/A	0.14	−0.02	7.7	5	2.8	−50	0.59	4.32
c5315	N/A	0.19	−0.01	N/A	0.67	0.08	N/A	0.29	−0.01	7.7	7.9	16.9	−50	−1.66	4.73
c6288	N/A	0.29	0.19	N/A	0	0	N/A	0.1	0.67	27.3	12.3	15.7	−50	10.43	10.21
c7552	N/A	0.28	−0.36	N/A	0.35	−0.05	N/A	0.56	1.77	16.7	9.3	15.7	−50	10.57	8.21
Average	N/A	0.4	0.31	N/A	0.48	0.14	N/A	0.19	0.33	9.2	10.7	15	−50	2	3.86

[a] Following the standard practice for 3D studies, Patnaik et al. report on area by considering individual die outlines. In [Pat+18e], area is reported in terms of die outlines as well

Moreover, she/he can identify all primary I/Os as they are implemented using wirebonds or TSVs, not randomized F2F vias.

2. *Layout hints:* Although the F2F vias are randomized, the attacker may try to correlate the proximity and orientation of F2F vias with their corresponding RDL connectivity. Toward this end, she/he can also investigate the routing toward the switchbox ports. Moreover, recalling the practical threat model, the attacker may be able to identify some known IP and confine the related sets of candidate F2F interconnects accordingly. The attack is generic and can account for those scenarios, by keeping track of the candidate F2F pairings considered by the attacker.

3. *Combinatorial loops:* Both tiers and thus all active components are available to the attacker, hence she/he can readily exclude those F2F connections inducing combinatorial loops.

The results in Table 6.6 indicate the efficiency of their proposed proximity attack (especially over the SAT-based attack [SRM15], see below for that scenario). Patnaik et al. assume that the attacker is able to infer all the driver-sink pairings for the switchboxes correctly; only the obfuscation within switchboxes remain to be attacked. In fact, this scenario can be considered as an optimal proximity attack, as for all F2F connections the correct one is always among the considered candidates. With regards to CCR, PNR, and HD for the recovered netlists, their protection scheme can be considered as reasonably secure (Fig. 6.15). Although PNR, which represents the degree of similarity between the original and the recovered netlist [Pat+18e], is around 30% or more for most benchmarks, HD approaches the ideal value of 50% for most benchmarks. In other words, although their attack can correctly recover some parts of the design, the overall functionality still remains obscured.

SAT-Based Attacks After manufacturing, the attacker can readily understand which four drivers/sinks are connected through the switchboxes, but she/he still has to resolve the obfuscation within the switchboxes themselves. The attacker may now leverage a working copy as an oracle and launch a SAT attack. Toward that end,

Table 6.6 Cut sizes and average attack runtimes. Time-out "t-o" is 72 h

Benchmark	Cut sizes		SAT attack [SRM15]	Proposed attack
	Random	Timing-aware	Runtime (min)	Runtime (s)
c432	134	56	624	0.004
c880	138	53	642	0.003
c1355	91	37	492	0.07
c3540	349	97	948	8.73
b17_1	6650	2482	t-o	0.25
b18	15,974	6906	t-o	113.51
b18_1	16,706	6616	t-o	4.62
b19	33,417	13,142	t-o	0.53

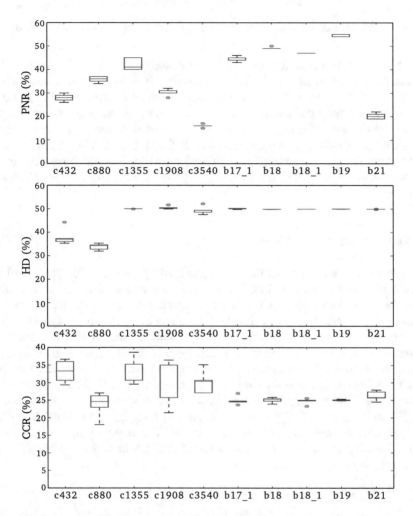

Fig. 6.15 Percentage of netlist recovery (PNR) [Pat+18e], Hamming distance (HD), and correct connection rates (CCR), when the benchmarks are subjected to their 3D split manufacturing proximity attack [Kne18]. Each box represents ten runs

Patnaik et al. employ the attack proposed in [SRM15], and they model the problem using multiplexers. Empirical results are given in Table 6.6. As expected, the SAT-based attack succeeds for smaller designs but runs into time-out for larger designs. This finding is also consistent with those reported by Xie et al. [XBS17] for their security-driven 2.5D scheme, which has a security notion similar to this case study.

6.4.4   Methodology for Prevention of Targeted Trojan Insertion

As elucidated in the introduction of the case study, aside from the need for IP protection, insertion of HTs by an untrustworthy foundry is another concerning threat. Security schemes like split manufacturing can hinder those adversaries from obtaining and fully understanding the netlist; hence, the adversaries may fail to insert HTs at particular targeted locations. However, there are many other parties in an IC supply chain which may leak the netlist to those fab-based adversaries. Therefore, following a strong threat model [Ime+13b, Li+18a] (reviewed next), Patnaik et al. leverage 3D integration as well to hinder such an advanced HT threat scenario.

6.4.4.1   Strong Threat Model

The security guarantee of *k-security*, as proposed by Imeson et al. [Ime+13b] is as follows. Given a *k-secure* FEOL layout and the complete, final gate-level netlist (just before splitting into FEOL/BEOL), an attacker has only a chance of $1/k$ for successful HT insertion into a particular location (or an up to k times higher risk for having $\leq k$ HTs detected by subsequent inspection). To achieve this, the idea is to induce k isomorphic structures in the FEOL by carefully lifting wires to the BEOL. As a result, an attacker cannot uniquely map these k structures to the target in the netlist, but can only randomly guess with a probability of $1/k$ (Fig. 6.16). Imeson et al. [Ime+13b] developed a greedy heuristic to select wires to be lifted to the BEOL, and also apply SAT to compute the security level, i.e., the minimal degree of isomorphism for any cell type found in the whole FEOL layout. Further discussions on *k-security*, its shortcomings, and advancements are included in Appendix for interested readers.

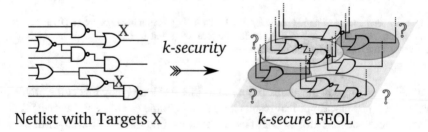

Netlist with Targets X   **k-secure FEOL**

Fig. 6.16 *k-security* involves applying split manufacturing and purposefully lifting wires to the BEOL (indicated by dashed wires) such that FEOL-based attackers cannot uniquely identify some or any parts of the netlist available to them. Hence, targeted Trojan insertion becomes difficult. Here, with security level two, an attacker has a 50% chance each for correctly inserting Trojans into the targeted OR and NOR-NAND structures

Note that there are two key pillars for this case study: (1) a security-driven synthesis strategy and (2) an end-to-end CAD flow for preventing HTs in 3D ICs. The motivation for a security-driven synthesis strategy is that [Ime+13b, Li+18a] implement their protection on top of a given netlist, solely as an afterthought. In contrast, by delegating the construction of isomorphic structures to the synthesis stage, Patnaik et al. effectively render the protection against HTs a design-time priority. Besides, their end-to-end CAD flow for 3D ICs, extended from the earlier part of this case study, effectively raises the notion of *k-security* toward practical application for preventing HT insertion in large-scale designs.

6.4.4.2   Security-Driven Synthesis Stage

The essence of security-driven synthesis strategy proposed by Patnaik et al. is as follows. For any netlist, one may assume that the designer can identify the structures vulnerable to HT insertion (e.g., by vulnerability analysis [Li+18a, ST16]) and wants to protect them accordingly. In agreement with *k-security*, the designer then intends to induce many isomorphic instances of those structures in the FEOL. To take control of the layout cost, but also to advance scalability and the attainable level of security, the authors delegate this step of inducing isomorphic instances to the synthesis stage. This way, their approach can be considered as "secure by construction." In this context, note that, although the authors do modify the netlist, they still assume—in agreement with [Ime+13b, Li+18a]—that the attacker holds the final gate-level netlist, truthfully representing their security-driven synthesis stage. As a result, the authors do not imply security through obscurity. Further details on the synthesis stage are provided in Fig. 6.17 and below.

Based on some vulnerability analysis of choice [Li+18a, ST16], the designer first identifies the vulnerable gates/structures. In this case study, Patnaik et al. leverage [Li+18a] to identify various structures which are covering the vulnerable gates as well as some surrounding gates (Fig. 6.18). It is understood that the designer can investigate as many structures as desired regarding (1) layout cost, (2) the potential for inducing isomorphic instances, and (3) the coverage of vulnerable gates. In fact, Patnaik et al. explored in total 18 structures; the ones illustrated in Fig. 6.18 are the most promising ones for their empirical study. See Sect. 6.4.5.1 for the layout cost of all the 18 structures investigated.

Next, these structures are prepared for synthesis, i.e., they are defined as custom cells. Typically, one has to conduct library characterization when creating custom cells. However, in this case, the structures are decomposed again later on, i.e., they are transformed back into their corresponding arrangement of simple two-input cells. It is essential, however, to track and preserve all gates related to the decomposed structures, by setting them as "don't touch," such that the design tools cannot interfere with these gates. Patnaik et al. also keep track of the input/output wires of the structures, which have to be lifted later on to the BEOL to achieve *k-security*. Now, instead of library characterization, the authors leverage the characteristics of simple cells available in the library (e.g., NAND with the same

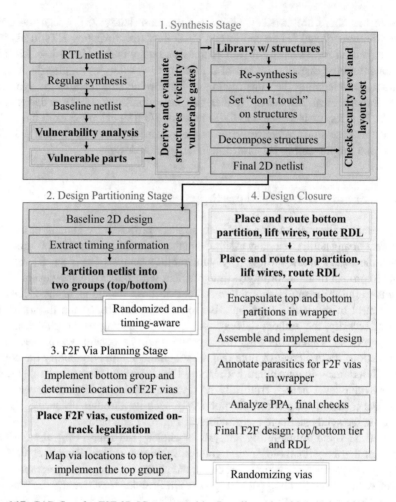

Fig. 6.17 CAD flow for F2F 3D ICs, proposed by Patnaik et al., with security-driven steps for preventing HT insertion emphasized in bold

number of inputs as the structure), but adapt the Boolean functionality as needed. The authors note that utilizing the characteristics of simple cells saves not only effort but also "tricks" the synthesis tool into using more instances for those structures (see also Sect. 6.4.5.1, Table 6.7). That is presumably because the structures provide some complex Boolean functionality with little, indeed over-optimistic layout cost. With their iterative synthesis approach, Patnaik et al. can thus impose more instances of the various structures as needed, while also correctly gauging the anticipated layout cost (by decomposing the structures).

Once the synthesis iterations are completed, which depends on the designer's considerations on the security level and/or synthesis-level PPA cost, the vulnerabil-

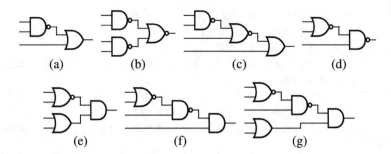

Fig. 6.18 The seven structures considered in this case study, without loss of generality. They are constructed based on the vulnerability analysis of [Li+18a]

Table 6.7 Isomorphic instances of structures as in Fig. 6.18, netlist coverage, and final security level *k* for the iterative re-synthesis. Bold-face represents the minimum k for each benchmark in an iteration

Benchmark	Iteration 1							Iteration 5						
	(a)	(b)	(c)	(d)	(e)	(f)	(g)	(a)	(b)	(c)	(d)	(e)	(f)	(g)
b14	89	39	16	79	151	29	**1**	137	43	**30**	164	179	65	33
b15_1	319	106	**37**	93	178	73	117	365	112	**46**	139	206	80	138
b17_1	1140	309	**127**	318	485	161	388	1239	333	**164**	424	532	207	442
b18	2170	778	**411**	612	1097	487	952	2583	870	**576**	1046	1243	712	1156
b19	5011	1558	**939**	1320	2422	1174	1669	5809	1745	**1221**	2420	2832	1762	2073
b20	164	85	**27**	206	208	86	0	308	99	**70**	425	277	180	72
b22	249	85	47	345	340	149	**7**	453	**108**	114	727	454	325	124

Benchmark	Design coverage, security level after iteration 5				
	(a)–(g)	Gates covered	Total gates	Coverage	k
b14	651	1780	5061	35.17%	30
b15_1	1086	3018	7639	39.51%	46
b17_1	3341	9173	23,912	38.36%	164
b18	8186	23,373	67,775	34.49%	576
b19	17,862	50,413	140,176	35.96%	1221
b20	1431	3882	11,116	34.92%	70
b22	2305	6298	16,941	37.18%	108

ity of the final netlist is to be re-evaluated. The final security level hinges on how many vulnerable gates are covered by the isomorphic instances of all structures.

6.4.4.3   End-to-End CAD Flow

As indicated, the 3D IC CAD flow for preventing targeted HT insertion is extended from the earlier part of this case study. However, there are some differences as follows.

1. Since the threat of HT insertion applies exclusively to manufacturing time, layout camouflaging/obfuscation is not required for the RDL. However, randomization of the routing paths within the RDL is required.
2. *k-security* is applied initially, to the whole netlist at once, using the security-driven synthesis strategy.
3. The flow proposed by Patnaik et al. comprises techniques similar to those required for regular split manufacturing. More specifically, the authors leverage their customized lifting cells [Pat+18e] to lift wires to the RDL as dictated by *k-security*.

The flow enables the concerned designer to tackle both layout cost and resilience against HT insertion. To do so, Patnaik et al. integrate their security-driven synthesis strategy as a key stage for the overall flow (Fig. 6.17). Further steps in their flow include partitioning, planning of F2F vias, and placement and split manufacturing-aware routing, whose details are provided below.

Design Partitioning After conducting the security-driven synthesis strategy, Patnaik et al. place and route the 2D design. The resulting netlist is partitioned into two groups, representing the top and bottom tiers of the F2F 3D IC. The authors define I/O ports for wires crossing the two tiers after partitioning; these ports become F2F vias later on.

As for the impact of 3D partitioning on *k-security*, it is important to note the following. First, Patnaik et al. ensure that all the gates of any decomposed isomorphic instance stay together in one tier. Second, since the authors apply the synthesis strategy on the complete netlist, i.e., before partitioning, there is no inherent limitation for isomorphic instances. Third, partitioning itself may, however, still impact the final security level. That is because once an attacker can retake the same partitioning steps as the designer, the attacker can also infer which subset of isomorphic instances goes into which tier. It is easy to see that fully random partitioning would render this attacker's benefit void, but the authors found that this can impose considerable layout cost.

Based on Sect. 6.4.2.1, Patnaik et al. propose a customized timing-driven and security-aware partitioning technique as follows. First, the authors obtain the timing reports for the 2D baseline layout. Then, each critical timing path without any isomorphic instances is kept within one tier. Other paths, i.e., paths with some isomorphic instances or non-critical paths, are randomly partitioned across the two tiers. With this partitioning technique, the attacker cannot understand which isomorphic instances in the bottom/top tier relate to which in the netlist—the definition of the security level as in [Ime+13b] is maintained.

Planning of F2F Interconnects These steps are primarily the same as in Sect. 6.4.2.2. In particular, the randomization of F2F ports and the custom on-track legalization are the same, albeit without the obfuscated switchboxes.

Wire Lifting and Design Closure Next, both tiers are placed and routed separately. In the absence of the obfuscated switchboxes leveraged for routing the RDL in Sect. 6.4.2.3, Patnaik et al. re-tailored their custom lifting cells [Pat+18e] to enable wire-lifting as required for *k-security*. That is, while routing the top/bottom tier, the authors route the regular metal layers, lift wires to the RDL with the help of the lifting cells, and route the RDL. For design closure, the top and bottom tiers are wrapped into one netlist. Again, the authors purposefully do not engage in any optimization across tiers, to maintain anonymized layouts. As in Sect. 6.4.2.3, the authors derive the SPEF from the wrapper netlist to capture the RC parasitics of F2F vias, and the final layout costs are evaluated.

6.4.5   Results for Prevention of Targeted Trojan Insertion

The setup is the same as in Sect. 6.4.3, except that two M8 layers are used for the RDL. That is also because without the need for obfuscated switchboxes, the RDL is less complex.

6.4.5.1   Analysis of the Security-Driven Synthesis Stage

During the iterative, security-driven synthesis stage, note that Patnaik et al. fix/preserve all gates of any isomorphic structure. The authors observe that doing so helps to guide the logical synthesis toward the remaining parts of the netlist not yet covered by some structures; The authors can increase the instance counts within a reasonable runtime. Since the synthesis iterations require only a few minutes for all the commonly considered benchmarks (e.g., 25–45 min even for *ITC-99*), the authors additionally explore the large-scale *IBM superblue* benchmarks for scalability of their synthesis strategy (Fig. 6.19). Here the authors observe that

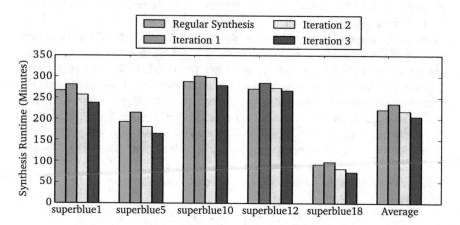

Fig. 6.19 Runtime comparison for regular synthesis and security-driven synthesis strategy for large-scale *IBM superblue* benchmarks

their strategy still impose only little runtime cost, about 6.3% on average for the first iteration, and runtimes for successive iterations are further reducing.

In Table 6.7, Patnaik et al. report on isomorphic instances and their coverage for various benchmarks, based on 10% of all gates being identified as vulnerable ones [Li+18a]. The authors find that large parts of the netlists can be covered by isomorphic instances already after a few synthesis iterations, namely 36.5%, on average, after five iterations. This coverage provides strong protection beyond the 10% of gates targeted. For the large-scale *IBM superblue* benchmarks (not illustrated in Table 6.7) the coverage is even higher, at 59%, on average. Although Patnaik et al. seek to cover all the gates which have been initially identified as vulnerable ones, some of those gates may not be covered in the final netlist, due to the iterative re-synthesis runs. Patnaik et al. found that this effect is acceptable; of the 10% identified gates, there are 9% covered on average. Besides, the authors note that few of the critical paths contain isomorphic instances to begin with. That is because critical paths rarely contain vulnerable gates [Li+18a, ST16]; related Trojans would be easy to detect by delay testing.

As for security levels following the definition in [Ime+13b], Patnaik et al. can achieve significant levels already after the fifth iteration, ranging from 30 for *ITC-99* benchmarks *b14* up to 1221 for *b19*. Based on the counts of individual instances, one may also revisit the synthesis after ruling out some structures which tend to limit the security levels, e.g., structure (c) for Table 6.7. Hence, the authors' strategy provides margin to the designer for both cases when lower levels are sufficient or higher levels are desired.

In Table 6.8, Patnaik et al. report on the impact of some synthesis iterations and the final 2D layout cost. For the latter, wire-lifting toward the BEOL as required by *k-security* is already accounted for. Overall, layout costs are acceptable; that is especially true when qualitatively comparing to [Ime+13b], where already notably smaller benchmarks induced significant PPA cost. That is also because the authors can safely apply buffer insertion to tackle timing degradation, as can be seen below.

As foreclosed, Patnaik et al. also provide the distribution of layout cost for all 18 different structures they explored in Figs. 6.20, 6.21, and 6.22. There the authors contrast the re-synthesized netlists without any optimization to those after buffer insertion. Since all isomorphic instances are preserved, buffer insertion cannot interfere with those structures and security is not undermined. This simple technique helps to avoid large cost while "tricking" synthesis into using the custom cells; it forms the baseline for the remaining, 3D-centric experiments in Sect. 6.4.5.2.

Finally, since the work by Patnaik et al. is the first to consider large-scale benchmarks for *k-security*, a direct comparison with prior works [Ime+13b, Li+18a] is impractical. For a qualitative comparison, the work by Patnaik et al. allows for superior security levels, induces little layout cost, and is scalable, all by means of synthesis.

Table 6.8 Gate counts and associated layout cost. Final layout-level cost account for wire-lifting (WL) in 2D. All cost are in % and with respect to the 2D baseline. A refers to area, P to power, and D to delay

Benchmark	Iteration 1				Iteration 2				Iteration 3			
	Gates	A	P	D	Gates	A	P	D	Gates	A	P	D
b14	4471	8.66	13.65	5.03	4683	12.78	19.11	5.10	4938	18.6	26.6	5.4
b15_1	7535	11.23	8.17	7.7	7622	12.67	10.95	7.58	7640	13.33	12.87	7.6
b17_1	23,491	13.21	11.67	7.55	23,876	14.94	14.07	8.05	23,960	15.00	15.57	7.54
b18	64,417	6.51	−10.87	17.64	66,222	9.32	3.47	17.57	67,134	10.01	4.04	14.74
b19	134,634	0.47	−29.9	24.27	137,478	2.73	−15.36	21.75	139,032	3.02	−13.94	20.61
b20	9716	4.94	4.32	3.33	10,268	9.61	11.68	3.31	10,835	15.13	17.44	3.66
b22	15,058	4.93	7.88	6.23	16,015	10.13	13.29	7.13	16,613	13.92	18.21	7.98

Benchmark	Iteration 4				Iteration 5				Final (with WL)		
	Gates	A	P	D	Gates	A	P	D	A	P	D
b14	4967	19.72	27.7	5.14	5061	21.62	28.64	8.75	21.62	62.36	55.65
b15_1	7658	14.39	13.12	7.65	7722	16.25	15.23	7.84	16.25	43.24	61.65
b17_1	24,250	15.58	19.15	9.03	25,112	18.95	23.18	15.5	93.95	58.96	63.54
b18	67,671	11.68	7.06	16.08	67,775	12.66	8.14	15.67	112.66	41.48	75.18
b19	140,016	3.71	−12.81	21.80	140,176	4.59	−13.3	21.37	154.59	51.17	88.87
b20	10,981	16.61	16.25	3.47	11,116	18.0	18.81	7.0	18.0	54.23	63.32
b22	16,941	15.96	20.17	10.6	16,963	16.14	21.72	10.72	16.14	54.44	66.61

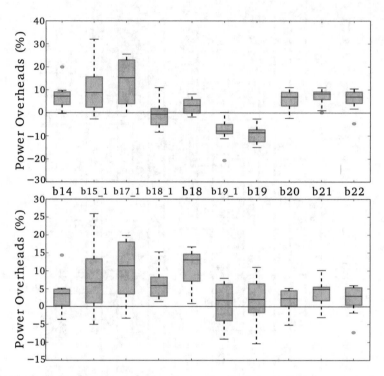

Fig. 6.20 Distribution of power cost for all 18 different structures Patnaik et al. explored on various benchmarks. Top: without any optimization; bottom: with buffer insertion applied (only outside of the structures). Note the different scales for each plot. Each boxplot represents ten runs

6.4.5.2 3D IC Layout Cost and Security Analysis

In Table 6.9, Patnaik et al. report on the final layout cost for the F2F 3D ICs. Overall, costs are better than for the 2D setup (Table 6.8), especially for larger benchmarks. This key finding attests their objective to advance *k-security* for large-scale designs. In Fig. 6.23, Patnaik et al. showcase the layouts for *ITC-99* benchmark *b18*.

Patnaik et al. also report on security levels in Table 6.9. The levels in 3D are the same as in 2D—as explained above, their 3D partitioning does not undermine security. In general, few if any proper attacks on *k-security* are available yet, and this is because the underlying notion of *k-isomorphism* is formally secure [CFL10]. This also implies that otherwise effective attacks will not be applicable. For example, although one can tackle the missing RDL connections using a SAT attack as [SRM15]—the netlist available to the adversary can serve as "virtual oracle" here—doing so is not practical for the following. First, as Patnaik et al. show in Table 6.9, SAT attacks become computationally expensive for large designs. Second, while a SAT attack may eventually provide a functionally equivalent assignment for the RDL connectivity, they cannot provide the structurally equivalent assignment required for attacking *k-security* any better than random guessing would.

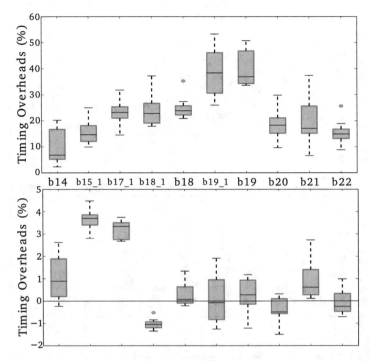

Fig. 6.21 Distribution of performance cost for all 18 different structures Patnaik et al. explored on various benchmarks. Top: without any optimization; bottom: with buffer insertion applied (only outside of the structures). Note the different scales for each plot. Each boxplot represents ten runs

6.5 Case Study II: 2.5D Integration for Secure System-Level Operation of Untrusted Chiplets

There are various kinds of hardware security features (HWSFs) offering some means of data protection, like the industrial *ARM TrustZone* and *Intel SGX* or the academic *MIT Sanctum* (these and others are reviewed in [Mae+18]). Notwithstanding their good prospects, most, if not all, HWSFs eventually become prone to tailored attacks. For example, Lee et al. [Lee+17] successfully exploited a memory corruption vulnerability for *Intel SGX* enclave, or Qiu et al. [Qiu+19] manipulate, exclusively at the software level, the supply voltages to induce security-violating hardware faults in *ARM TrustZone*.

Furthermore, for adversaries throughout the IC supply chain, HWSFs may form prime targets—strategic or collaborative adversaries would first aim to bypass or disable the HWSFs so that attacks conducted in the field can remain "under the radar" thereafter. Arguably, none of the proposed HWSFs in the prior art can fully

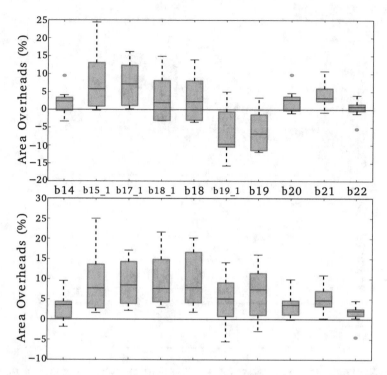

Fig. 6.22 Distribution of area cost for all 18 different structures Patnaik et al. explored on various benchmarks. Top: without any optimization; bottom: with buffer insertion applied (only outside of the structures). Note the different scales for each plot. Each boxplot represents ten runs

Table 6.9 Isomorphic instances, security level, layout cost (in %) with respect to 2D baseline, and SAT attacks (t-o is 100 h)

	Structures		Security level[a]	Overheads			SAT attack [SRM15]
Benchmark	Bottom	Top	k	Area	Power	Delay	Runtime
b14	16	14	30	−40.32	30.66	37.17	t-o
b15_1	27	19	46	−40.21	27.69	38.32	t-o
b17_1	84	80	164	−37.92	46.56	45.73	t-o
b18	200	145	345	−29.39	22.16	51.09	t-o
b19	200	200	400	−25.48	8.46	62.44	t-o
b20	38	32	70	−38.32	42.46	47.86	t-o
b22	58	50	108	−34.47	45.52	46.92	t-o

[a] The level is defined as the sum of the least occurring structures in the bottom and top tier; see also the security-aware partitioning technique in Sect. 6.4.4.3

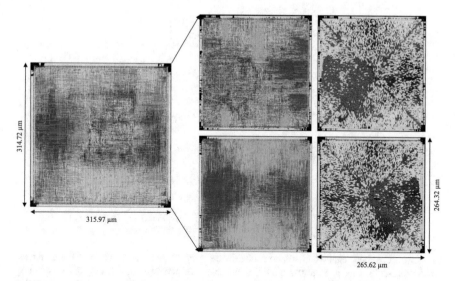

Fig. 6.23 2D layout (left), bottom/top tier of the F2F 3D IC (right), both for benchmark b18. Yellow dots (very right) illustrate the F2F vias

withstand malicious modifications.[1] The above outlined circumstances also impose an important practical challenge, namely how to implement ICs that are high-end, competitive, and relatively cheap, yet trustworthy and secure (Fig. 6.24). Few, if any, of the outsourced high-end facilities may be considered trustworthy, whereas maintaining an advanced, trusted facility on-shore or in-house is too costly in practice.

In this case study, we describe the work by Nabeel et al. [Nab+20] in detail. The key idea of their work is to integrate untrustworthy commodity chiplets using a security-enforcing interposer. Such hardware organization provides a robust 2.5D root of trust (RoT) for trustworthy, yet powerful and flexible, computation systems. The scope of this case study can be summarized as follows:

- The 2.5D RoT concept establishes stringent physical separation at the system level, between (1) commodity chiplets and (2) HWSFs residing in an active interposer. In addition to ruling out common threat scenarios directly by construction, the purpose of this concept is to enable continuous runtime monitoring of the system-level communication of all commodity chiplets.

---

[1] There are efforts to render hardware secure in the direct presence of Trojans; such schemes typically leverage some formalism like multi-party computation [Bro+18] or verification and proofing [Wah+16]. While promising, such schemes still require that at least some parts of the system remain trustworthy, i.e., that some parts are guaranteed to be free of any malicious modifications. Moreover, such schemes are less applicable to general-purpose, high-performance computation systems, as the underlying formalism requires extensive system- and circuit-level support, which naturally also tends to impose considerable overheads.

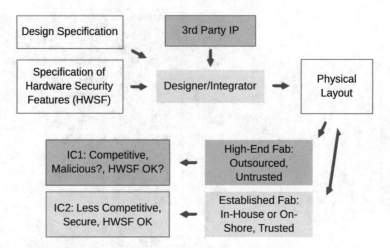

Fig. 6.24 IC supply chain with focus on hardware security features (HWSFs). Green and red boxes represent trusted and untrusted entities and assets, respectively. Implementing a trustworthy and competitive IC requires a trusted and high-end fabrication process, two aspects that conflict with each other, as also indicated in [BRS17, Mav+17]

- The 2.5D RoT concept is showcased on a secure multi-core architecture with a system-level interconnect fabric and shared memories. Dedicated HWSFs for memory access and data control are developed for the same. The license-free parts of this proof-of-concept (PoC) 64-core implementation are made available to the community [Nab20].
- An end-to-end physical-design flow to enable the 2.5D RoT concept is developed on commercial tools. The proposed flow serves to design the active interposer and supports a flexible design mode for chiplets procured as soft or hard IP. Using this novel flow, the layout costs of the scheme are elaborated in detail.
- The 2.5D RoT scheme is evaluated against various relevant attack scenarios. That is, the related security-enforcing policies are implemented and demonstrated against malicious runtime behavior, using a commercial hardware simulation workflow.

6.5.1   Concepts and Background for the 2.5D Root of Trust

The first step in leveraging 2.5D integration to advance security is to study the integration options and the related implications for security and practicality. Thereafter, in the subsequent subsections, this case study covers the details on a secure-by-construction 2.5D architecture for runtime monitoring along with its implementation, as well as its evaluation (the latter in terms of security as well as layout costs).

6.5.1.1   2.5D Integration and Implications for Security and Practicality

Nabeel et al. harness the opportunities offered by state-of-the-art 2.5D technologies for advancing hardware security. Recall that the 2.5D interposer technology in general is becoming more and more relevant and practical (Sect. 6.3). More specifically, Nabeel et al. propose the assembly of: (1) potentially untrusted commodity chiplets and memories, and (2) physically separated, entrusted communication interfaces and HWSFs residing in an active interposer. Nabeel et al. refer to the resulting system in general and the security-enforcing interposer in particular as 2.5D RoT (Fig. 6.25).

From a commercial point of view, note that the system vendor has to design, produce, and sell such 2.5D RoT systems. Here, the good economics of chiplets reuse are still maintained; that vendor has to manufacture only a fraction of the overall system, namely the security-enforcing interposer (with the help of some established and trusted fabrication facilities) and would then integrate the high-end but untrusted chiplets on top of that interposer (with the help of in-house or certified on-shore packaging facilities). Therefore, the final system establishes security and can offer good performance at a reasonable cost—the crux illustrated in Fig. 6.24 can thus be resolved.

At this point, one might wonder about using a passive interposer for an alternative, potentially less costly implementation of an 2.5D RoT, but Nabeel et al. argue that doing so would entail two key limitations. First, scalability would be compromised. This is because all HWSFs would have to be implemented within one or multiple, dedicated security chiplet(s), which would then become the "bottleneck" for system-level communication through the interposer. In fact, optimizing interposer interconnects is an area of research by itself, where active interposers are considered promising as well [Akg+16, Yin+18]. Second, the need

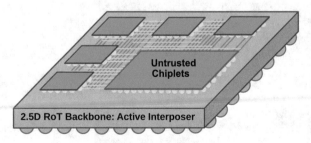

Fig. 6.25 An active interposer acts as backbone for the 2.5D root of trust (RoT), which enables secure system-level integration of untrusted chiplets. All system-level communication is "policed" by the interposer

for trustworthy manufacturing of an interposer and some security chiplet(s) might well undermine the good economics of the scheme. In short, Nabeel et al. advocate for an active interposer for their proposed 2.5D RoT.

6.5.1.2   Threat Model

This work does not require any trust assurance concerning the design and manufacturing of commodity chiplets. In fact, Nabeel et al. even assume a priori that chiplets do run malicious code and/or incorporate Trojans.

Crucially, such threats cannot undermine or compromise the system-level security of this scheme. This is due to the fact that the 2.5D RoT imposes physically and inevitably that any untrusted component has to depend on the security-enforcing interposer for system-level communication, whereas the trustworthiness and robust operation of that interposer are not subject to those untrusted components. Note that this is in contrast to most prior art where HWSFs are embedded monolithically in the same chip and, thus, remain subject to the trustworthiness—or rather lack thereof—of all the related design and manufacturing stages.

The concept and its threat model are illustrated in Fig. 6.26, with details for both discussed next.

When seeking to securely integrate various components at the system level, different threats are to be considered, which concern the system-level communication and all involved components [Bas+17]. More specifically, in this work, a malicious chiplet may exercise the following attacks:

1. Passive reading, also known as snooping, i.e., a chiplet illicitly reads or gathers data that is meant for/authorized to other chiplets;
2. Masquerading, also known as spoofing, i.e., a chiplet disguises or poses itself as another one, to illicitly control services or request data from other chiplets;
3. Modifying, i.e., a chiplet maliciously changes the data exchanged legally between other chiplets;

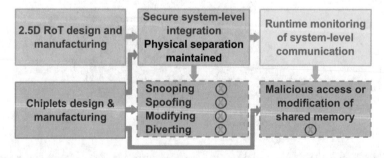

Fig. 6.26 2.5D RoT design and manufacturing stages (green), threat sources (red), and runtime monitoring (turquoise). It is imperative that the active interposer is designed and manufactured by trusted parties

4. Diverting, i.e., a chiplet maliciously diverts the data exchanged legally between two chiplets to a third, unauthorized chiplet; and/or
5. Man-in-the-middle, i.e., a chiplet "hijacks" the communication between two chiplets—this attack is closely related to all four above.

As Nabeel et al. focus on a multi-core architecture with shared memories, they also have to consider another threat:

6. Malicious accesses and modifications of shared-memory-resident data.

It is assumed that any of these six threats can be introduced by: (a) untrusted components/chiplets—either unintentionally via "design bugs" or intentionally via Trojans—or (b) malicious software running on the cores.

It is assumed that any attack is exercised through system-level communication across chiplets. Therefore, any adversarial activities conducted within chiplets, such as covert channels across cores (e.g., [Mas+15]), side-channel- or fault-driven attacks across cores and their caches/buffers (e.g., [OST05, Sch+19]), or fault injection on privileged hardware interfaces (e.g., [Mur+20]) are all considered out of scope for this work. Furthermore, a trusted runtime environment is assumed. Thus, any threats like side-channel or physical fault-injection attacks conducted by malicious end-users (e.g., [BCO04]) are considered out of scope as well.

Finally, it is assumed that the design and fabrication of commodity chiplets is outsourced and, hence, untrusted, whereas the design and manufacturing of the 2.5D RoT and the system-level assembly must all be carried out in a trusted environment. This also means that Nabeel et al. do not seek to detect or prevent Trojans within chiplets; recall that they rather assume a priori that Trojans are present in the untrusted chiplets. Given that related techniques (e.g., [LM19, Cha+09b]) are orthogonal to the efforts of this, such techniques could still be leveraged, to render the final system even more robust to begin with.

Note that, among other scenarios, *DARPA's CHIPS* and *DARPA's Secure Processing Architecture by Design (SPADE)* programs both match well with said assumptions. This is because government agencies seeking to build small numbers of large-scale, heterogeneous systems in a cost-efficient manner are advised to utilize chiplets which, when obtained from the open market, are potentially malicious. To ensure secure computation nevertheless, within a trusted runtime environment readily enforceable by the agencies, schemes like this presented here become essential.

6.5.1.3   Working Principles

An important novelty of this work is that it rules out the above threats (1)–(5), all by design and construction. As Fig. 6.27 illustrates, this kind of built-in security occurs as the system-level interconnect fabric with all its interfaces and HWSFs are physically separated from the untrusted components. Therefore, components/chiplets remain completely unaware of and isolated from any communication not directly

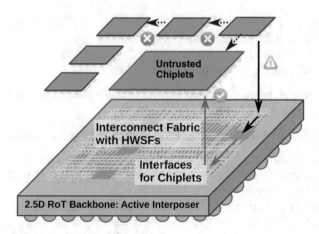

Fig. 6.27 The scheme prevents common threats like snooping by construction. System-level communication must utilize the 2.5D RoT backbone and, thus, cannot be tampered with by other chiplets (red crosses). This is because the interposer's interfaces, which each chiplet is physically attached to, enforce that any communication request (yellow warning sign) is passed to and controlled by the HWSFs residing in the active interposer. Only approved communication (green tick) is passed on

addressed to or created by them. For example, regarding spoofing, Nabeel et al. realize a hard-coded assignment of component identifiers (IDs) directly via the interconnect interfaces which are residing exclusively in the 2.5D RoT. Thus, a malicious component cannot masquerade itself as another in the first place.

Furthermore, Nabeel et al. utilize the notion of security policies for runtime monitoring against malicious access or modification of the system-level shared memory. To do so, the authors devise HWSFs that allow them to enforce a fully controlled memory access scheme. Nabeel et al. apply stringent principles as follows, with related technical details provided in Sect. 6.5.2.

First, any memory access not explicitly allowed for, via some policy, is denied by default. Second, the continuous "policing" of memory access incurs uniform latency, independent of whether access is allowed or denied, and any denied access is responded to with a generic error message. These principles in conjunction ensure that an adversary cannot infer whether the requested region is protected or not used at all, which may serve well to hinder any related side-channel inference. Third, to protect against faults and malicious data modifications within memories themselves, we advocate for optional memory-security features like error correction codes (ECCs). Here, the actual data and the results of the security features are to be stored in physically separate locations and cross-checked upon reading.

In short, common threats are ruled out by construction, malicious memory accesses are blocked, and erroneous data is rejected; all these security principles are enabled directly at the 2.5D RoT. For any such adversarial case, the overall system may experience a (temporary) loss of functionality or data, but its integrity and trustworthiness remain intact, an outcome which constitutes the main focus of this work.

6.5.2   Architecture and Implementation of 2.5D Root of Trust

First, the architecture of the 2.5D RoT proposed by Nabeel et al., called Interposer-based Security-Enforcing Architecture (ISEA) is described. Thereafter, its implementation is outlined in detail. The key paradigms of ISEA are: (1) to physically separate commodity components (chiplets in this work) from the HWSFs and (2) to monitor any memory-related, system-level communication at runtime.

The novelty and enabler for ISEA is the security-enforcing active interposer, which serves as integration carrier and as "physical barrier" for any communication-centric security fallacies to propagate through the system. Toward that end, the interposer hosts the system-level interconnect fabric along with all proposed HWSFs. Therefore, any communication emanating from untrusted chiplets is inevitably handled and controlled by the interposer. More specifically, Nabeel et al. focus on shared-memory transactions initiated by cores residing within chiplets. The legality of any such transaction is verified using various kinds of security policies; details and examples for such policies are provided further below and in Sect. 6.5.3.2.

6.5.2.1   System Implementation

Overview

Figure 6.28 depicts the block diagram for the proposed ISEA architecture. Key to ISEA are Transaction Monitors (TRANSMONs) which administer the various policies; the functionalities and implementation of TRANSMONs and all other HWSFs are explained in detail further below.

Without loss of generality, Nabeel et al. consider the *ARM Cortex-M0* core for the commodity chiplets. For the system-level interconnect fabric, the authors leverage the *ARM Advanced High-Performance Bus Lite* (AHB-Lite). AHB-Lite facilitates communication among bus-attached master components (cores in this work) that initiate transactions and bus-attached slave components (system-level shared memories in this work) that respond to these requests. AHB-Lite transfers data values, addresses, and control info; it is managed by components such as arbiters, decoders, multiplexers, etc., all of which collectively implement the Advanced Microcontroller Bus Architecture (AMBA) protocol. Nabeel et al. choose AHB-Lite as it is technology-independent, widely used in the industry, and encourages modular design, all while offering high performance. Note that AMBA provides a secondary bus which functions as a slave to AHB-Lite, called the Advanced Peripheral Bus (APB), used for lower-bandwidth peripheral devices such as I/O

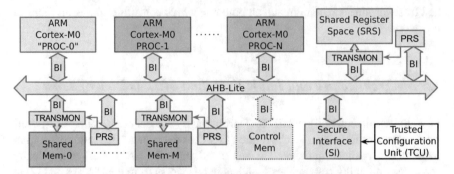

Fig. 6.28 Block diagram for the proposed ISEA architecture. Components shown in green are security-critical and implemented exclusively in the 2.5D RoT; components shown in red constitute untrusted commodity components, implemented in various chiplets. Note that all the bus interfaces (BIs) are also implemented in the 2.5D RoT; this serves to ensure that master IDs for any transaction emanating from some core cannot be tampered with by any of the other cores. Also note that "PROC-0" is a trustworthy core which is separate from all the cores in the commodity chiplets; it is implemented entirely in the 2.5D RoT and reserved for scheduling and other management tasks such as compilation and updating of security policies residing in the PRSs (policy register spaces). For the optional memory-security features (residing in TRANSMONs, not illustrated separately here), the related control memory is to be provided as a separate, trusted memory chiplet

ports. While APB could also be incorporated into ISEA, the authors focus on AHB-Lite in this work.

Nabeel et al. release the license-free parts of their PoC implementation to the community [Nab20]. The authors exhibit only a particular instance of ISEA here; their scheme can be easily retro-fitted to secure other systems, with different chiplets, cores, and/or interconnects. This is because the key principles of the scheme are agnostic to these implementation aspects.

Note that Cortex-M0 does not provide a cross-communication interface; hence, direct M0-to-M0 communication is not possible, thereby also excluding such direct message-passing for this PoC implementation of ISEA. Still, ISEA could be extended toward direct message-passing; TRANSMONs would be incorporated directly in-between the components to be monitored. This approach would also be applicable for, e.g., traditional system-on-chip (SoC) designs where IP cores may be connected via direct links. Besides, modern architectures may also contain hardware accelerators which can use integrated memories and/or external, shared memories. For the latter, the accelerators have to act as a bus master, like all other cores, to access those shared memories. Thus, ISEA can also be used to monitor transactions by other components, like accelerators, not just regular cores. Finally, ISEA could also be extended toward other types of system-level fabrics, like NoCs.

It should be emphasized again that it is essential for ISEA that the system-level interconnect fabric, its interfaces, and all proposed HWSFs are implemented exclusively in the active interposer, thereby constituting the 2.5D RoT by design and construction. For example, all communication requests passed onto the AHB-

Lite bus system are associated with a master ID, whose assignment is handled by the bus-interface ports the chiplets are physically attached to—for ISEA, these ports are implemented in the trusted interposer, not within the chiplet. Thus, concerning spoofing, by construction there is no attack surface that could be leveraged by some Trojan or malicious software running within the chiplets and seeking to alter the master IDs. This and other scenarios are also illustrated in Fig. 6.33, Sect. 6.5.3.2.

Hardware Security Features of ISEA

1. TRANSMONs, along with their Policy Register Spaces (PRS) to store the various policies;
2. A Shared Register Space (SRS);
3. An ARM Cortex-M0 core called "PROC-0"; and
4. The Secure Interface (SI).

The purpose of these features and their components is explained next.

(1) A TRANSMON controls all transactions related to its attached memory chiplet, based on the policies stored in its PRS. A TRANSMON itself comprises three or four components, namely the Address Protection Unit (APU), the Data Protection Unit (DPU), the Slave Access Filter (SAF), and, optionally, a memory-security feature. All components establish security collectively, with their functionality elaborated in Sect. 6.5.2.2. In a nutshell, the APU protects against undefined and/or unpermitted memory accesses, the DPU protects against illegal data modification or leakage of restricted data into the system's shared-memory space, the SAF serves to forward or reject requests which are approved or rejected, respectively, and the memory-security feature serves to detect faults and malicious modifications within the memories themselves.

(2) The SRS can be used for secure data sharing, e.g., for semaphores. Although the SRS is implemented in the interposer, a TRANSMON and its related PRS are still required, to realize access control and runtime monitoring.

(3) The interposer-embedded (and thus fully trustworthy) PROC-0 serves for scheduling and controlling the distributed computation, with commodity cores in the untrusted chiplets being allocated and interrupted by PROC-0 at runtime as needed. PROC-0 will further serve for mapping the system-level shared-memory spaces, and for compiling and updating the application-specific sets of policies residing in the PRSs. It is important to note that PROC-0 does not constitute a "bottleneck" as it is not involved in each and every AHB-Lite transaction, but it is only used in exercising this kind of system-level management.

(4) An external Trusted Configuration Unit (TCU) is responsible for loading the application(s) and initial data onto the system, and for retrieving the final results from the system. All these tasks are performed using the SI, which has privileged access to the AHB-Lite. Recall that a trusted runtime environment is assumed; attacks misusing the TCU or SI in the field are thus out of scope. In any case, access to the TCU or SI can be protected by cryptographic primitives.

6.5.2.2   Transaction Monitor

The key to ISEA's operation is TRANSMONs, with their micro-architecture illustrated in Fig. 6.29. Note that an individual TRANSMON is placed in-between every memory slave and the AHB-Lite bus interface (Fig. 6.28). While another option would be to place TRANSMONs in-between all the core masters and their respective bus interfaces, this design decision offers two important benefits. First, a TRANSMON connected to a master would require additional address bits decoding and checking (for the base address), which is already covered by AHB-Lite itself, whereas a TRANSMON connected to a slave only requires decoding and checking for the offset address. Second, a TRANSMON connected to a slave allows to keep track of the security policies relevant to only that slave, thereby helping with efficiency.

TRANSMON Design: Overview, Working Principles As described, a TRANS-MON comprises an APU, a DPU, an SAF, optionally a memory-security feature, and some glue logic. The APU and DPU each have access to their own PRS. For efficiency, every PRS is implemented using flip-flops. Each APU PRS entry defines one APU policy concerning some particular region in the system's shared-memory space, physically allocated in the memory slave connected to that TRANSMON; each DPU PRS entry defines one DPU policy concerning some particular data. Both APU and DPU policies are discussed in more detail below and examples are illustrated in Figs. 6.30 and 6.31.

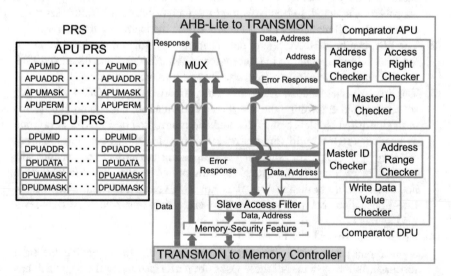

Fig. 6.29 Block diagram for the TRANSMON's micro-architecture. See also Sect. 6.5.2.2 for technical details as well as Figs. 6.30 and 6.31 for examples of APU, DPU policies in action, respectively

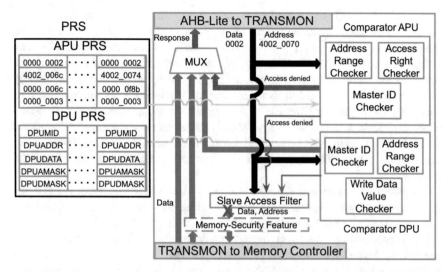

Fig. 6.30 APU policies in action in the TRANSMON. Essentially, the core with ID 0x2 tries to access a memory region outside of the allowed ranges defined in the two policies. Therefore, the access is blocked by the SAF, Slave Access Filter, from passing to the memory controller, and an error message is returned. See also Sect. 6.5.3.2 for more details

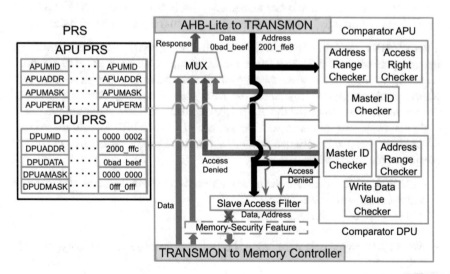

Fig. 6.31 A DPU policy in action in the TRANSMON. Essentially, the core with ID 0x2 tries to write out a sensitive software asset, e.g., a private cryptographic key, with the value of 0xBAD_BEEF, to a restricted memory region. The access is blocked by the DPU from passing to the memory controller, and an error message is returned. See also Sect. 6.5.3.2 for more details

TRANSMONs block all read or write requests that are violating any of their APU/DPU policies. By default, TRANSMONs also block requests that cannot be matched to any policy, protecting the system against all such "stray requests." Policy verification also involves the checking of master/slave IDs. In this context, as the interconnect fabric and all its interfaces are physically implemented exclusively in the active interposer, recall that there cannot, a priori, be any spoofing of IDs, snooping, modifying, or diverting of data, or man-in-the-middle attacks. In case a request is blocked, the related TRANSMON passes an error message to the master which initiated the transaction and an interrupt to the trusted PROC-0. The memory access itself is then dropped by the SAF—it is thus guaranteed to never reach the memory.

As indicated, the trusted PROC-0 within ISEA serves for mapping the system-level shared-memory spaces, and for compiling and updating the application-specific sets of policies for each TRANSMON (more specifically, for its PRS). For different applications running on the system, depending on the scheduling, policies can also be devised for protection of independent data sets of multiple applications running in parallel. Once a particular application run is finalized, before dropping the related policies, PROC-0 should also clear the related memory regions, to avoid any posterior leakage of sensitive data. Moreover, the trusted end-user is free to implement software-level analysis and management of all blocked requests. Such management schemes may also decide whether masters, which repetitively trigger requests to be blocked should be isolated completely from the system (by updating the policies accordingly), in order to mitigate potential denial-of-service attacks.

TRANSMON Design: Address Protection Unit (APU) The APU forms an integral part of the TRANSMON; it serves to check all read or write memory requests. As such, full access control over all shared-memory ranges is exercised. Nabeel et al. design and implement the APU such that policy checking is acting during the address phase of the AHB-Lite protocol, thereby avoiding additional cycle delays. Recall that each APU makes use of its own PRS to hold the polices related to its physically assigned memory slave. As Fig. 6.29 shows, an APU policy comprises four parameters:

- APUMID, which identifies the master allowed to initiate the particular memory request described by this policy;
- APUADDR, a 32-bit memory address;
- APUMASK, a 32-bit address mask; and
- APUPERM, the access permission, i.e., whether read-only, write-only, or read-write.

An example for an APU policy is illustrated in Fig. 6.30, and the related simulation is provided in Sect. 6.5.3.2.

TRANSMON Design: Data Protection Unit (DPU) The DPU forms another integral part of the TRANSMON, and its function is to provide data-level protection. This is achieved by blocking: (1) over-writing of sensitive data in the event of unauthorized writes to specific memory locations, or (2) writing out particular data of sensitive nature. The latter serves to protect *soft assets*, e.g., private cryptographic keys, from leaking inadvertently into the system's shared memory, e.g., by malicious "shadow writes" [Bas+17].

A write transaction is blocked when the DPU PRS contains a relevant policy that disables writing of particular, restricted data to a specified address range. Since DPU policy checks can only work during the data phase of the AHB-Lite protocol, one has to keep the data, address, and control signals all registered until the check is completed; this registering is done within the SAF. Hence, the DPU incurs one additional cycle delay in all transactions related to write-restricted data, but all other transactions not covered by DPU policies are not delayed.

As with the APU, recall that a DPU makes use of its own PRS. Figure 6.29 shows the five parameters of a DPU policy:

- DPUMID, which identifies the master whose write transaction is to be verified against this policy;
- DPUADDR, a 32-bit memory address, which designates the address where the write permission is restricted;
- DPUDATA, a 32-bit, write-restricted data value;
- DPUDMASK, a 32-bit data mask; and
- DPUAMASK, a 32-bit address mask.

An example for a DPU policy is illustrated in Fig. 6.31, and the related simulation is provided in Sect. 6.5.3.2.

TRANSMON Design: Memory-Security Feature To protect against faults or malicious modifications within the shared system-level memories themselves, schemes like ECC, cyclic redundancy check (CRC), data mirroring, or a combination of these can be implemented. For example, an ECC implementation based on the well-known Hamming code would require four extra bits per memory byte, translating to 50% memory cost, and could only serve to detect at most two corrupted bits per byte. The advantage of ECC, however, is that it can be calculated during computation time without any latency overhead. A CRC implementation is more suitable when the memory data to protect is not supposed to change, e.g., for a firmware/software image. Also, CRC can be implemented with little additional circuitry. Still, the CRC computation needs to be run for chunks of data intermittently; this may halt some regular computation and can thus impact the overall throughput. In data mirroring, the data is simply copied into another (trusted) memory, which would naturally induce an overhead of 100%.

Nabeel et al. envision some memory-security scheme as follows: (1) the TRANSMON computes an ECC for any write out that is allowed; (2) the ECC result is stored in some separate and trusted control memory, whereas the actual data is stored in the shared-memory chiplet attached to the TRANSMON; (3) during read-out, the TRANSMON validates the data using the stored ECC result. If that check fails and cannot be corrected via the ECC, the data is rejected and the related memory region is marked as tainted and not used further.

Since address handling is covered by the AHB-Lite protocol, one may implement this scheme such that ECC results are fetched in parallel, without inducing additional delays for read-out transactions. We note that ECC results have to be stored in a trusted, separate memory chiplet. Finally, neither the above nor other memory-security features can protect against erroneous data arising from hardware/software failures or malicious activities. Such risks can only be mitigated at the system level, e.g., by redundant computation and majority voting on results [Mav+17, Ngu+08]. Note that a multi-core architecture like the one presented here can be readily tailored for such needs.

6.5.2.3 Physical Design

Next, the end-to-end physical-design flow, as devised for enabling the 2.5D RoT by Nabeel et al. is presented. Note that this flow can also be applied for any other active 2.5D system. The flow is illustrated in Fig. 6.32, and some highlights are discussed next. Note that the flow is leveraging commercial tools, libraries, and technologies for all key design steps such as placement and routing or handling of timing constraints; see also Sect. 6.5.3.1 for more details on the setup.

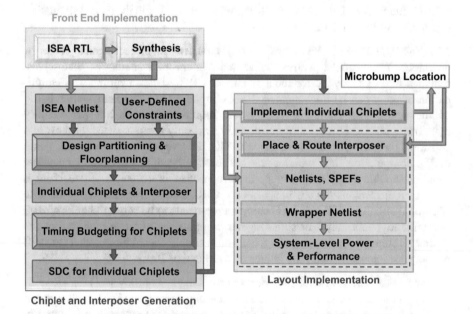

Fig. 6.32 End-to-end physical-design flow based on commercial tools. After synthesis of ISEA, the chiplets and interposer are floorplanned. Next, all components are implemented, along with microbumps planning. If chiplets are given as hard IPs, only the interposer implementation is required (dashed box)

First, for the front-end implementation, the whole ISEA register-transfer level (RTL) design is synthesized to obtain a full-system netlist. Chiplets provided as soft IP are to be synthesized here as well.

Chiplets and Interposer Generation The full-system netlist is then partitioned into banks, which simply represent the logic and memory chiplets. The flow provides flexibility to the designer when choosing the number of logic/memory banks as needed for the chiplets organization. Based on the full-system netlist, the timing budgets are derived and separate timing constraints (SDC files) are obtained for the individual chiplets and the active interposer.

Next, the full-system floorplan is generated. Relevant parameters are to be provided by the designer, such as utilization for individual chiplets and their aspect ratio, and they are used toward floorplanning of the related core/memory banks. Also, the floorplans of memory banks capture the placement of memory modules within each bank. The designer is also required to provide the arrangement of chiplets over the active interposer. Finally, the interposer die outline is derived from the full-system floorplan.

All the floorplan data is kept in *Tool Command Language (TCL)* format, which eases the use of a regular 2D implementation flow while designing the chiplets and the active interposer. Nabeel et al. emphasize that their flow is flexible with respect to accommodating chiplets that are either designed in-house or, what is more practical, procured as physical hard IP from commercial vendors. For such hard IP, the design steps are more straightforward and essentially cover only the chiplets arrangement over the interposer and the design of the active interposer itself. When procuring such hard IP, it is easy to see that the designer has no freedom for any intra-chiplet optimization. Still, the flow allows the designer to explore different chiplet arrangements, which eases the system-level design space exploration along with an investigation of timing and power consumption.

Layout Implementation For chiplets obtained as soft IP, the related netlists have to proceed through a standard 2D implementation flow, to obtain the individually placed-and-routed chiplet layouts. During this step, the locations for microbumps, which serve the physical connection between chiplets and the interposer, are also derived. Those microbumps are initially placed around the vicinity of drivers/sinks, while further on-track legalization is performed to avoid routability issues and maximize the utilization of routing resources for the chiplets. Thereafter, the microbump locations of all chiplets are used to define the microbump locations for the interposer. Next, the RC parasitics for each chiplet are generated as SPEF files from their post-routed layouts. Along with the final netlist, these SPEF files are used later on for sign-off analysis, i.e., to evaluate power consumption and timing.

Once the 2D implementation of all chiplets is completed—which is skipped in case chiplets are obtained as hard IP—the placement and routing of the active interposer follow. First, the interposer netlist is imported, which describes the AHB-Lite components, the HWSFs of ISEA, and the pre-defined interposer microbump locations. Second, a 2D implementation of the active interposer follows. Note that Nabeel et al. do not engage in any cross-optimization between chiplets and interposer, which is essential for the scenario of chiplets obtained as hard IP. Third, the RC parasitics for the active interposer design are extracted and exported along with the final netlist, and the GDSII is streamed out. Finally, the RC parasitics for the microbumps are modeled into the SPEF file of a wrapper netlist.

To evaluate the system-level power consumption and timing of ISEA, including all computing and memory chiplets, all individual netlists and their SPEF files are used along the wrapper netlist with its own SPEF file.

6.5.3   Experimental Evaluation

6.5.3.1   Setup

The RTL code for the complete system, including the cores, AHB-Lite bus, TRANSMONs, etc., has been realized using *Verilog*. Nabeel et al. release the license-free parts of the RTL [Nab20]. Synthesis was performed via *Synopsys DC* and layout generation via *Cadence Innovus v.17.10*. Verification and simulation runs have been carried out via *Synopsys VCS*. The *ARM IAR* suite has been used to compile *C* code to run on ISEA.

Nabeel et al. implement ISEA as 64-core *ARM Cortex-M0* multi-chiplet system for a PoC. As baseline, the 64 cores are organized into four computing chiplets, each holding 16 cores. For another configuration, to study the impact of system organization on layout costs, the authors reorganize the 64 cores into eight chiplets, each holding eight cores. Concerning security policies, the baseline configuration supports 16 APU and 16 DPU policies for each TRANSMON. To study the impact of policies being supported by TRANSMONs on layout costs, the authors also consider configurations with 32, 64, and 128 APU and DPU policies being supported by each TRANSMON.

For both the computing and shared-memory chiplets, Nabeel et al. leverage the commercial 65 nm *GlobalFoundries* technology and *ARM* standard cell and memory libraries, representing the advanced but untrusted facility. The authors employ four shared-memory chiplets with 1 MB SRAM each, build up from 16 memories at 64 kB. For the active interposer, the authors use the *Synopsys SAED* 90 nm technology, representing the older but trusted facility. For brevity, both technologies are also referred to as 65 and 90 nm, respectively. Note that the 90 nm technology does not provide memory modules; thus, the authors had to refrain from implementing any memory-security feature for this PoC, as they cannot provision for a separate, trusted memory chiplet required for such features. For both technologies, the authors use a supply voltage of 1.08 V and consider their respective slow corners. Note that doing so allows for heterogeneous 2.5D integration without the need for level shifters. In reality, the advanced but untrusted facility versus the older but trusted facility may support technology nodes that are further apart, but the authors were constrained in choices by the libraries available to them. Microbumps connecting the interposer and chiplets have a width of 5 μm and a pitch of 10 μm. The authors utilize seven metal layers for both the 90 nm and the 65 nm technology.

6.5.3.2   Security Analysis

Nabeel et al. study various scenarios for securing computation using ISEA. First, they illustrate how critical threats (i.e., snooping, spoofing, modifying, diverting, and man-in-the-middle attacks) are ruled out by ISEA in the first place (Fig. 6.33). More specifically, there is an approved transaction between PROC-2 and the shared memory with slave ID 0 (represented as blue arrow and green check in the TRANSMON of the memory). At the same time, PROC-1 seeks to snoop on that communication. This threat is blocked physically, directly by the BI (bus interface) of PROC-1, as the BI itself delegates only data originating from/destined to PROC-1. Next, PROC-1 tries to illicitly act as man-in-the-middle between PROC-2 and the shared memory with slave ID 4 (represented as dashed, dark-red arrow). This threat is blocked directly at the BI as well—the BI hard-codes the master ID 1 into any outgoing request, thereby preventing PROC-1 from masquerading its ID. Finally, PROC-1 also tries to access some data in the shared memory with slave ID 0 (orange arrow). However, this particular request is not approved by any policy and, thus, rejected. PROC-0 is informed about this blocked request as well (black arrow).

Next, Nabeel et al. explore various scenarios for runtime monitoring against malicious access or modification of the system-level shared memory. These scenarios serve to show-case the working of ISEA in some detail, based on hardware simulation using *Synopsys VCS* along with *C* code compiled for the Cortex-

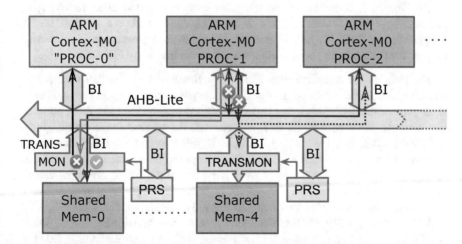

Fig. 6.33 Various scenarios for system-level communication as handled within ISEA. Red crosses mean that the related threats are prevented, whereas a green check means that the transaction is approved

Table 6.10 Selected signals for AHB-lite

Signal	Description
HCLK	Bus clock; timing of all signals is related to the rising edge of HCLK
HMASTER	Master ID; a unique ID assigned to each master attached to the bus
HSEL	Slave select; indicates that the current transaction is intended for the selected slave
HADDR	System address; identifies the address as related to the slave
HWDATA	Write data; used to transfer data from the master to the bus slaves during write operations and vice versa for read operations
HWRITE	Transfer direction; HIGH indicates a write transfer, whereas LOW indicates a read transfer
HREADY	Transfer status; HIGH indicates that a transfer has finished on the bus; to extend the transaction, this signal is to be driven LOW
HRESP	Transfer response; provides feedback on the status of the transfer; used as receipt for security approval/rejection in this work

M0 cores using *ARM IAR*. AHB-Lite signals relevant for understanding of these simulations are listed in Table 6.10. Aside from the particular scenarios considered, the number of policies can increase and their interaction can become more complex in practice, depending on the application(s) running on ISEA and the resulting security requirements.

Protection of Memory Ranges Here, ISEA is tasked exemplarily with executing a fast Fourier transformation (FFT). The FFT is an essential building block for many signal processing applications, and it can be parallelized straightforwardly. As indicated in Sect. 6.5.2, task scheduling is handled by the trusted control processor PROC-0, which also arranges the input data within the system's shared-memory space. The FFT computations within each core are started upon receiving an interrupt from PROC-0, and once the processing is done for all cores, the final results are gathered by PROC-0. Toward that end, Nabeel et al. implement custom interrupt handler for the cores, to perform computation as controlled by PROC-0.

The policies are compiled such that the intermediate FFT results calculated by one core cannot be modified by other, maliciously acting cores. That is, Nabeel et al. protect the shared-memory regions assigned to each core via APU policies. For example, the core with ID 0x2 has access to the address range 0x4002_0000 to 0x4002_006C and the range from 0x4002_0074 to 0x4002_0FFF, but not to other addresses such as 0x4002_0070 (i.e., where the core with ID 0x1 stores its result). Note that the address ranges are derived by the APU in an efficient manner, i.e., without need for complex comparator logic, using simple bit-wise operations. For example for the related simulation in Fig. 6.34, the start address 0x4002_0000 is *APUADDR[1] AND NOT(APUMASK[1])* and the end address 0x4002_006C is *APUADDR[1] OR APUMASK[1]*; similarly, the start address 0x4002_0074 is *APUADDR[2] AND NOT(APUMASK[2])* and the end address 0x4002_0FFF is *APUADDR[2] OR APUMASK[2]*.

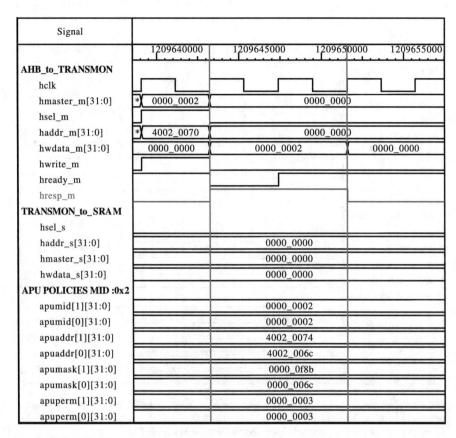

Fig. 6.34 Demonstration of ISEA at hardware simulation level, using *Synopsys VCS*. Malicious manipulation of data is blocked by an APU policy. The blocking is receipted by the error message *hresp_m*. AHB-Lite signals are described in Table 6.10

As shown in the waveform in Fig. 6.34, the core with ID 0x2 tries to access the address 0x4002_0070 to write out the data 0x0000_0002. Note that for AHB-Lite in general, the address phase comes first and the data phase one cycle after. The transaction is blocked by the APU, and the data in the memory remain protected and as is, indicated by the fact that the memory-controller signals are *not* reflecting the requested write out. At the same time, the error message *hresp_m* is returned. Finally, note that this particular example is the same as in Fig. 6.30.

Protection of Private Assets Here, a malicious core tries to write out some soft asset, e.g., a private cryptographic key. The DPU covers this kind of threat; the related DPU policy concerns the actual data.

For Fig. 6.35, a DPU policy is set to track a write transaction by the core with ID 0x2 to the restricted memory region between addresses 0x2000_0000 to 0x2FFF_FFFF, concerning the sensitive data 0x0BAD_BEEF. Note that the

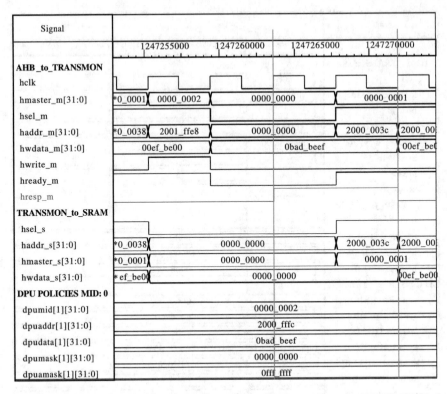

Fig. 6.35 Demonstration of ISEA at hardware simulation level, using *Synopsys VCS*. Unpermitted write out of a secret key *0XBAD_BEEF* is blocked by a DPU policy. The blocking is receipted by the error message *hresp_m*. AHB-Lite signals are described in Table 6.10

DPU derives address ranges like the APU; the start address is *DPUADDR AND NOT(DPUAMASK)* and the end address is *DPUADDR OR DPUAMASK*. Also, the sensitive data is derived similarly, as *DPUDATA AND NOT(DPUMASK)*.

The simulation waveform in Fig. 6.35 shows an attempt to write out the restricted data value to address 0x2001_FFE8, which is blocked. Here as well, the error message *hresp_m* is returned. Note that subsequently another, unrelated read transaction is approved, which can be seen by the *hready_m* signal being turned on during the related data phase. Finally, note that this particular example is the same as in the conceptional Fig. 6.31.

Protection of Shared Assets Here, it is assumed that two or more cores require a semaphore for software-based program and data control. Semaphores can be stored in the SRS, the shared register space, which is part of the 2.5D RoT, hence trustworthy by itself (Sect. 6.5.2). Consider a maliciously acting core tries to overwrite the semaphore to be able to access/execute data/program regions otherwise not accessible. Here, a DPU policy is needed to monitor the actual data access to the semaphore, whereas a generic APU policy would not suffice.

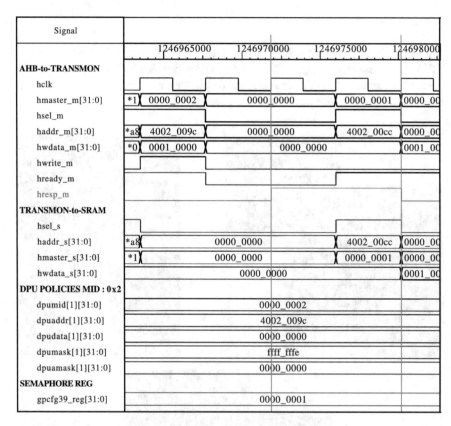

Signal	1246965000	1246970000	1246975000	1246980000
AHB-to-TRANSMON				
hclk				
hmaster_m[31:0]	*1〉 0000_0002 〉	0000_0000	〉 0000_0001 〉	0000_00
hsel_m				
haddr_m[31:0]	*a8〉 4002_009c 〉	0000_0000	〉 4002_00cc 〉	0000_00
hwdata_m[31:0]	*0〉 0001_0000 〉	0000_0000		〉 0001_00
hwrite_m				
hready_m				
hresp_m				
TRANSMON-to-SRAM				
hsel_s				
haddr_s[31:0]	*a8〉	0000_0000	〉 4002_00cc 〉	0000_00
hmaster_s[31:0]	*1〉	0000_0000	〉 0000_0001 〉	0000_00
hwdata_s[31:0]		0000_0000		〉 0001_00
DPU POLICIES MID : 0x2				
dpumid[1][31:0]		0000_0002		
dpuaddr[1][31:0]		4002_009c		
dpudata[1][31:0]		0000_0000		
dpumask[1][31:0]		ffff_fffe		
dpuamask[1][31:0]		0000_0000		
SEMAPHORE REG				
gpcfg39_reg[31:0]		0000_0001		

Fig. 6.36 Demonstration of ISEA at hardware simulation level, using *Synopsys VCS*. Malicious over-writing of a semaphore in *gpcfg39_reg* is blocked by a DPU policy. The blocking is receipted by the error message *hresp_m*. AHB-Lite signals are described in Table 6.10

Figure 6.36 shows how such a malicious transaction is blocked. Here *gpcfg39_reg* is considered as a semaphore register. For the core with ID 0x1, to obtain the ownership of this semaphore, it has to write 0x0000_0001 to the above register, but can do so only while the semaphore register value is 0x0000_0000, i.e., while the semaphore is available. For the core with ID 0x2, it has to write 0x0000_0010 to obtain the semaphore, and so on. Naturally, one core should not be able to obtain the semaphore when it is already used by any other core—a DPU policy is compiled to implement this restriction. In the simulation, the policy is set for the core with ID 0x02 to prevent any malicious writing of "0" to the last bit (*DPUDATA AND NOT(DPUMASK)*) of the semaphore register. The waveform shows such an attempt to clear that last bit, which is blocked, along with the error message *hresp_m* being returned. Note that the waveform shows subsequently another, unrelated read transaction initiated by the core with ID 0x01, which is approved.

6.5.3.3   Layout Analysis

Using their 2.5D design flow, Nabeel et al. investigate the physical layouts of various ISEA configurations. In Fig. 6.37, the authors provide snapshots for the baseline 64-core 2.5D version of ISEA. Note that the details discussed below are based on the commercial-grade implementation setup (Sect. 6.5.3.1).

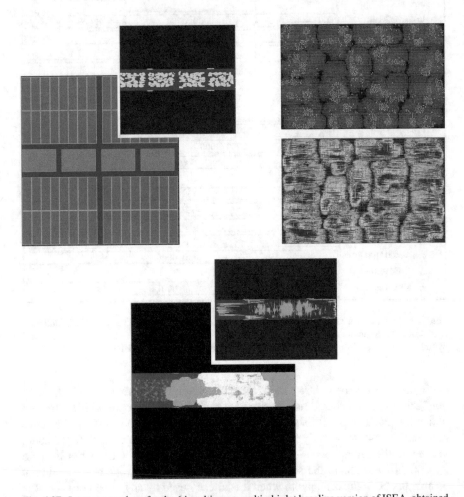

Fig. 6.37 Layout snapshots for the 64 multi-core, multi-chiplet baseline version of ISEA, obtained using the 2.5D design flow. For visual clarity, the power-distribution networks are not shown for the routing snapshots. Top left: floorplan of ISEA. Four ARM chiplets in the middle and four shared-memory chiplets around; Top left, inset: interposer microbump locations. Top right: One ARM Cortex-M0 chiplet, with 16 cores. Logic and microbump locations (top), routing (bottom). Bottom: ISEA implementation in the active interposer. TRANSMON and other HWSFs in white, regular AHB-Lite components in grey. Placement regions are constrained, for cost-efficient interposers, e.g., using *Intel's EMIB* technology; Bottom, inset: interposer routing

2D Implementation, ISEA in General First, Nabeel et al. analyze the impact of their security-enforcing features on layout costs. To do so, the authors compare the baseline ISEA multi-core design to a corresponding but non-secure design, both implemented via a regular 2D IC flow using the *GlobalFoundries* 65 nm technology. For the non-secure design, the authors maintain all Cortex-M0 cores, memories, AHB-Lite components, and glue logic, but they drop all HWSFs such as TRANSMONs, PRSs, etc. Note that, from a conceptional point of view, using the *Synopsys SAED* 90 nm technology would be more apt, as this technology was designated as the trusted node. Then, the corresponding secure 2D implementation would represent the system as implemented exclusively using the trusted technology. However, given that the 90 nm technology does not provide memory modules, the authors have to resort to the 65 nm technology. For this reason, the authors also refrain from directly comparing the secure 2D implementation with the secure 2.5D system later on.

Table 6.11 provides the results for the 2D implementation. For the secure design, one can observe a 5% reduction in critical delay and a 13.86% increase in power consumption. Note that Nabeel et al. achieve a competitive critical delay for the secure design by breaking longer paths using pipelining. An increase in standard-cell area (2.48%), instance count (29.57%), buffer count (18.46%), wirelength (31.49%), and total capacitance (35.44%) are all expected, due to the proposed HWSFs (including all registers required for storage of policies, etc.) and due to pipelining. The die outline remains as is, however; no additional silicon cost occurs. These results provide the range of costs to be expected for ISEA, that is at least for this particular PoC implementation.

2.5D Implementation Table 6.12 provides the physical-design results for the 2.5D baseline implementations. As indicated, computing and memory chiplets are implemented using the 65 nm technology and the active interposer using the 90 nm technology, respectively. Here Nabeel et al. also compare a secure design with a non-secure design; both contain the same set of computing and memory chiplets, and both hold all AHB-Lite components in the active interposer, whereas the secure design further holds the proposed HWSFs in the interposer.

Table 6.11 2D Implementation results for non-secure versus secure designs, both in *GlobalFoundries* 65 nm

Metrics	Non-secure (2D)	Secure (2D)
Critical delay (ns)	9.79	9.29
Power consumption (mW)	239.5	272.7
Standard-cell area (μm^2)	24,127,403	24,725,036
Total die area (μm^2)	31,996,800	31,996,800
Total instance count	600,729	778,393
Total buffer count	132,477	156,929
Total wirelength (m)	28.9	38.2
Total capacitance (nF)	7.9	10.7

Table 6.12 2.5D Implementation results for non-secure versus secure designs, chiplets in *GlobalFoundries* 65 nm, Interposer in *Synopsys SAED* 90 nm

Metrics	Non-secure (2.5D)	Secure (2.5D)
Critical delay (ns)	9.72	9.83
Power consumption (mW)	266.4	300.9
Standard-cell area (μm^2)	24,588,292	26,844,473
Total die area (μm^2)	33,641,866	33,641,866
Interposer die area (μm^2)	6,237,600	6,237,600
Total instance count	569,574	745,693
Interposer instance count	69,742	249,085
Total buffer count	141,151	169,344
Total wirelength (m)	30.5	40.5
Total capacitance (nF)	7.92	10.89

For the secure design, Nabeel et al. observe an overhead of 1.13% for critical delay, 12.95% for power consumption, and 32.79% for wirelength, respectively. The standard-cell area is increased by 9.18%, while instance count and buffer counts are increased by 30.92% and 19.97%, respectively. As before, these costs are attributed to the HWSFs (including all PRS registers, etc.), but here the costs are further impacted by the migration to 2.5D and by the heterogeneous technology setup. More specifically, due to the migration to 2.5D, all the system-level interconnects are now passing through the active interposer, with all chiplets connected to this fabric through microbumps. Thus, timing closure for the interposer is subject to the multiple chiplets, which requires more effort. More importantly even, recall that the active interposer is implemented in the older 90 nm technology. Therefore, higher costs are naturally to be expected, especially for all the HWSFs residing in the interposer. As with the 2D designs, there is no impact on the die areas for the 2.5D designs. In fact, the size of the interposer is dominated by the size and arrangement of the chiplets mounted on top of it, not by the standard-cell area of the additional logic incurred for the HWSFs within the interposer.

It should be emphasized again that Nabeel et al. refrain from any cross-optimization between chiplets and the active interposer, to account for the practical assumption of hard-IP chiplets obtained as commodity components from the open market. Moreover, the authors note that their flow allows the designer to constrain the active area of the interposer (Fig. 6.37(bottom)) and the placement of microbumps (Fig. 6.37(top left)). Doing so enables the final vendor to manufacture only a small CMOS chip for the interposer, instead of the whole outline, which naturally helps save commercial cost. Such a small chip could be supported by Intel's EMIB technology [Int19].

Note that the results above are all subject to the ISEA PoC baseline configuration, i.e., 64 cores are organized into four computing chiplets, each holding 16 cores, and 16 APU and 16 DPU policies are supported by each TRANSMON. To understand the scaling of layout costs incurred by the proposed HWSFs, Nabeel et al. conduct the following experiments:

1. The author explore the impact for the number of policies being supported, by re-implementing the active interposer for 32, 64, and 128 APU and DPU policies being supported by each TRANSMON;
2. The authors explore the impact of the system-level organization, by rearranging the 64 cores into eight computing chiplets with eight cores each and re-implementing the whole system.

The results for (1) are illustrated in Fig. 6.38. Note that layout costs are scaling up as expected. More specifically, for each doubling of the number of policies being supported, most metrics are approximately doubled in cost as well (considering their respective baseline cost for the initial configuration supporting 16 policies). However, critical delays increase only linearly—this indicates that the physical designs are well optimized in terms of performance/timing paths.

The results for (2) are as follows: the standard-cell area is increased by 0.99%, interposer die area by 47.35%, power by 2.13%, critical delay by −10.15% (i.e., reduced by 10.15%), total instance count is increased by 6.24%, total wirelength by 9.41%, and total capacitance by 6.24%, respectively. Given the reorganization of cores into double the number of computing chiplets, such costs are expected. More specifically, on the one hand, having to accommodate double the computing chiplets imposes a larger outline for the interposer. This is because the chiplets are not halved in size, as microbumps are dominating their outlines, and not the logic within. Due to the larger interposer die outline, one can also observe larger total wirelength, along with more instances (required for buffering), higher capacitance,

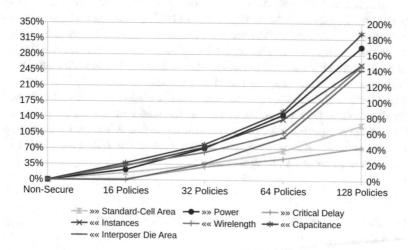

Fig. 6.38 Scaling of layout costs for the 2.5D RoT, for varying numbers of APU and DPU policies being supported by each TRANSMON. The legend indicates by "««" or "»»" the assignment of each metric to the left or the right y-axis, respectively. Each metric is normalized to its respective value obtained for the non-secure 2.5D design

and marginally higher power consumption. On the other hand, the critical delay can be improved, thanks to some critical paths becoming shorter within the smaller computing chiplets, as well as due to the rearrangement of chiplets on top of the active interposer.

In short, these two experiments show reasonable costs, manifesting the practicality of the architecture for different configurations. It can also be argued that, for other systems beyond the Cortex-M0 PoC implementation, e.g., when using *RISC-V* cores instead, the layout costs might be better amortized over the respectively larger system.

6.6 Closing Remarks

Heterogeneous 2.5D and 3D integration are technologies that have become more and more relevant in the industry, given their good prospects to advance the scale, ease of integration, as well as performance for modern electronics. In addition, unlike traditional 2D circuit integration, 2.5D and 3D integration open up, quite literally, another dimension to advance various aspects of hardware security. This chapter covered the technology background, concepts for advancing hardware security, as well as two case studies in some detail. The first case study demonstrated how 3D integration serves well to combine split manufacturing and camouflaging. In particular, the concept of "3D splitting," a design across multiple tiers along with randomization and camouflaging of the vertical interconnects between the tiers, was shown to enable a modern approach to (1) IP protection and (2) prevention of targeted Trojan insertion. The second case study highlighted how 2.5D integration allows a stringent physical separation, directly at the system level, between untrusted commodity components and trusted security-enforcing components.

This chapter aims to show the reader that leveraging 2.5D and 3D integration is indeed practical and promising to address security challenges that had been so far difficult to tackle. It also attempts to educate the reader on the further potential of 2.5D and 3D integration to advance security, by means of encapsulating sensitive circuitry within 3D ICs, or by extending the show-cased 2.5D root of trust with capabilities to monitor for physical attacks.

Appendix: Prior Art on k-Security, Its Limitations, and Advancements

Although it provides a formal foundation against HT insertion, *k-security* [Ime+13b] has practical shortcomings as follows.

"−" The security level hinges on the underlying composition of gates which, in turn, is dictated by logical synthesis. The fewer the instances of some gate type in the FEOL, the lower the overall security level. That is because *k-security* is defined as the minimum security level across all gate types. The technology library plays a key role here as well—the "richer" the library in terms of gate types, the lower the average counts for some individual types, and the lower the overall security level. Patnaik et al. report on this correlation in Table 6.13. In short, being agnostic to synthesis, [Ime+13b] is at the mercy of the design tools.

"−" The reported layout cost is considerable. For the *ISCAS-85* benchmark *c432*, for security level 8, the overheads are already 61%, 82%, and 207% for PPA, respectively. Hence, applying *k-security* on large designs can become prohibitive.

"−" Computing the security level is *NP-complete* and related to the problem of subgraph isomorphism. Imeson et al. [Ime+13b] leverage SAT solvers and circuit partitioning to limit the computational cost. Still, Patnaik et al. note that the authors did not investigate large-scale benchmarks with hundreds of thousands of gates. Also, while circuit partitioning is a common practice, here it hinders to explore the security for the whole netlist holistically, as sub-circuits can only be secured individually after partitioning.

"−" Imeson et al. [Ime+13b] delegate only wires but no gates to the trusted 2.5D facility. This approach is the same as with classical split manufacturing in 2D ICs, hence the complete potential of 3D integration is not utilized.

Li et al. [Li+18a] recently proposed an advancement for *k-security*, with the following contributions (labeled as "+"). Despite the advances proposed in [Li+18a], there are also some limitations (labeled as "−").

"+" Additional dummy gates (and dummy wires) help to raise the security level as required for layouts initially holding only a few instances of some gate type. To model the insertion of dummy gates and simultaneous lifting of wires, Li et al. [Li+18a] consider the concept of *spanning subgraph isomorphism*.

"+" To limit computational cost, they propose a mixed-integer linear program (MILP)-based framework.

"+" To limit layout cost, they protect only selected vulnerable gates in the layout, not the whole design. They further impose a close but uniform arrangement of isomorphic instances, to limit placement disturbances without leaking layout-level hints.

"−" As with Imeson et al. [Ime+13b], Li et al. [Li+18a] apply partitioning on larger netlists, effectively hindering to secure industrial designs in their entirety. Also, Li et al. [Li+18a] do not investigate any large-scale designs.

"−" While Li et al. [Li+18a] may raise the security level using dummy gates (and dummy wires), these gates can impose significant area overheads (and timing overheads, due to routing congestion when having to lift all related wires). Patnaik et al. report on the PPA impact for adding dummy gates/wires in Table 6.14. For the *ITC-99* benchmark *b17_1*, for security level 20, there are

Table 6.13 Gate composition dictating *k-security* level for various benchmarks under different libraries as in [Ime+13b]. All gates have driving strength X1. The maximal security levels equal the minimum count across the libraries' gate types (marked in boldface); the actual level may be lower

Benchmark	lib-3			lib-4				lib-5				
	NOR2	NAND2	INV	NOR2	NAND3	NAND2	INV	NOR3	NOR2	NAND3	NAND2	INV
c432	61	69	**30**	43	**14**	65	24	15	37	**9**	50	29
c7552	**461**	967	538	244	**158**	986	586	**36**	227	126	985	527
b14	**806**	3094	1282	450	**422**	2503	1092	**86**	370	362	2587	1151
b17_1	7097	12,909	**4206**	5549	**3179**	8500	4195	**1440**	4806	2462	7759	4454
b22	**2405**	10,818	4618	**1433**	1458	9129	4093	**223**	1398	1347	8599	3953

Benchmark	lib-7						
	NOR4	NOR3	NOR2	NAND4	NAND3	NAND2	INV
c432	**5**	9	45	**5**	**5**	37	31
c7552	**2**	7	230	59	179	940	523
b14	**7**	17	259	88	303	2598	1122
b1	**281**	585	3733	1070	1394	8533	4339
b22	**71**	92	1183	361	1033	8823	3825

Table 6.14 Study on [Li+18a], netlists provided as courtesy by the authors of [Li+18a]. Setup lib-3 is as in Table 6.13, and lib-8 contains NAND2, NOR2, AND2, OR2, XOR2, XNOR2, INV, and BUF, all in X1 strength. Number of gates and wires for original, unprotected layouts versus number of additional dummy gates (D. Gates) and lifted wires (L. Wires) for security levels 10 (S10) and 20 (S20) when 10% of original gates are considered vulnerable

Benchmark	lib-3 (Original)		lib-8 (Original)	
	Gates	Wires	Gates	Wires
b14	5182	5457	4125	4400
b15_1	7722	8207	6978	7463
b17_1	24,212	25,664	21,500	22,952
b20	11,810	12,332	10,686	9226
b22	17,841	18,608	14,457	15,224

Benchmark	lib-3 (S10)		lib-8 (S10)		lib-3 (S20)		lib-8 (S20)	
	D. Gates	L. Wires	D. Gates	L. Wires	D. Gates	L. Wires	D. Gates	L. Wires
b14	3620	12,600	4706	14,892	7819	25,012	9885	29,737
b15_1	6422	21,480	7185	22,205	13,708	43,124	15,246	44,306
b17_1	19,988	66,271	22,648	71,341	42,216	132,153	47,867	142,487
b20	8340	27,818	9748	33,661	17,808	55,602	22,405	67,044
b22	12,923	44,153	16,333	50,858	27,299	87,915	34,284	101,352

up to 1.2× more gates and 5.2× more wires when compared to the unprotected designs.[2]

"−" Li et al. do not utilize the potential of 3D integration.

In short, Imeson et al. [Ime+13b] and Li et al. [Li+18a] provide a solid formal foundation to protect against targeted HT insertion, but there are practical limitations to both schemes.

References

[AEM18] N.E.C. Akkaya, B. Erbagci, K. Ma, A secure camouflaged logic family using PostManufacturing programming with a 3.6GHz adder prototype in 65nm CMOS at 1V Nominal VDD, in *Proc. Int. Sol.-St. Circ. Conf.* (2018)

[Akg+16] I. Akgun et al., Scalable memory fabric for silicon interposer-based multi-core systems, in *Proc. Int. Conf. Comp. Des.* (2016), pp. 33–40. https://doi.org/10. 1109/ICCD.2016.7753258

---

[2] In [Ime+13b] it is only mandated that each vulnerable gate shall have $k - 1$ isomorphic instances, whereas Li et al. [Li+18a] further require that none of these instances is vulnerable themselves which, arguably, provides a more stringent security notion. Dummy gates/wires are necessary for this purpose, to compensate whenever some of the isomorphic instances are vulnerable themselves. In this case study referring to the work by Patnaik et al., the notion of [Ime+13b] is followed, while the methods can be easily tailored toward [Li+18a] as well.

[Bas+17] A. Basak et al., Security assurance for system-on-chip designs with untrusted IPs. Trans. Inf. Forens. Sec. **12**(7), 1515–1528 (2017). ISSN: 1556-6013. https://doi.org/10.1109/TIFS.2017.2658544

[BCO04] E. Brier, C. Clavier, F. Olivier, Correlation power analysis with a leakage model, in *Proc. Cryptogr. Hardw. Embed. Sys.* (2004)

[Bri+12a] S. Briais et al., 3D Hardware Canaries, in *Proc. Cryptogr. Hardw. Embed. Sys.*, ed. by E. Prouff, P. Schaumont (Springer, Berlin, Heidelberg, 2012), pp. 1–22. ISBN: 978-3-642-33027-8. https://doi.org/10.1007/978-3-642-33027-8_1

[Bro+18] O. Bronchain et al., Implementing trojan-resilient hardware from (mostly) untrusted components designed by colluding manufacturers, in *Proc. Attacks Sol. Hardw. Secur.* (2018), pp. 1–10. https://doi.org/10.1145/3266444.3266447

[BRS17] S. Bhunia, S. Ray, S. Sur-Kolay, eds. *Fundamentals of IP and SoC Security* (Springer, New York, 2017). https://doi.org/10.1007/978-3-319-50057-7

[BS15] C. Bao, A. Srivastava, 3D Integration: New opportunities in defense against cache-timing side-channel attacks, in *Proc. Int. Conf. Comp. Des.* (2015), pp. 273–280. https://doi.org/10.1109/ICCD.2015.7357114

[BS19] C. Bao, A. Srivastava, Reducing timing side-channel information leakage using 3D integration. Trans. Dependable Sec. Comput. **16**(4), 665–678 (2019). ISSN: 1545-5971. https://doi.org/10.1109/TDSC.2017.2712156

[CEP19] *Common Evaluation Platform.* MIT Lincoln Laboratory. 2019. https://github.com/mit-ll/CEP

[CFL10] J. Cheng, A. Wai-chee Fu, J. Liu, K-isomorphism: privacy preserving network publication against structural attacks, in *Proc. SIGMOD*. Indianapolis, Indiana (2010), pp. 459–470. ISBN: 978-1-4503-0032-2. https://doi.org/10.1145/1807167.1807218

[Cha+09b] R.S. Chakraborty et al., MERO: A statistical approach for hardware Trojan detection, in *Proc. Cryptogr. Hardw. Embed. Sys.* (2009), pp. 396–410

[Cha+16] K. Chang et al., Cascade2D: a design-aware partitioning approach to monolithic 3D IC with 2D commercial tools, in *Proc. Int. Conf. Comp.-Aided Des.* Austin, Texas (2016), 130:1–130:8. ISBN: 978-1-4503-4466-1. https://doi.org/10.1145/2966986.2967013

[Che+15] S. Chen et al., Chip-level anti-reverse engineering using transformable interconnects, in *Proc. Int. Symp. Def. Fault Tol. in VLSI Nanotech. Sys.* (2015), pp. 109–114. https://doi.org/10.1109/DFT.2015.7315145

[Cio+14] J. M. Cioranesco et al. "Cryptographically secure shields". In: *Proc. Int. Symp. Hardw.-Orient. Sec. Trust.* 2014, pp. 25–31. DOI: 10.1109/HST.2014.6855563.

[Cle+16] F. Clermidy et al., New perspectives for multicore architectures using advanced technologies, in *Proc. Int. Elec. Devices Meeting* (2016), pp. 35.1.1–35.1.4. https://doi.org/10.1109/IEDM.2016.7838545

[Dof+16] J. Dofe et al., Hardware security threats and potential countermeasures in emerging 3D ICs, in *Proc. Great Lakes Symp. VLSI.* Boston, Massachusetts (2016), pp. 69–74. ISBN: 978-1-4503-4274-2. https://doi.org/10.1145/2902961.2903014

[Dof+17] J. Dofe et al., Security threats and countermeasures in three-dimensional integrated circuits, in *Proc. Great Lakes Symp. VLSI.* Banff, Alberta (2017), pp. 321–326. ISBN: 978-1-4503-4972-7. https://doi.org/10.1145/3060403.3060500

[DRR17] J. DeVale, R. Rakvic, K. Rudd, Another dimension in integrated circuit trust. J. Cryptogr. Eng. **8**(4), 315–326 (2017). ISSN: 2190-8516. https://doi.org/10.1007/s13389-017-0164-7

[DY18] J. Dofe, Q. Yu, Exploiting PDN noise to thwart correlation power analysis attacks in 3D ICs, in *Proc. Int. Worksh. Sys.-Level Interconn. Pred.* (2018). https://doi.org/10.1145/3225209.3225212

[Fic+13] D. Fick et al., Centip3De: a cluster-based NTC architecture with 64 ARM cortex-M3 cores in 3D stacked 130 nm CMOS. J. Sol. State. Circ. **48**(1), 104–117 (2013). ISSN: 0018-9200. https://doi.org/10.1109/JSSC.2012.2222814

[GM13] S. Garg, D. Marculescu, Mitigating the impact of process variation on the performance of 3-D integrated circuits. Trans. VLSI Syst. **21**(10), 1903–1914 (2013). ISSN: 1063–8210. https://doi.org/10.1109/TVLSI.2012.2226762

[Gre16] D.S. Green. Common heterogeneous integration and IP reuse strategies (CHIPS). DARPA. 2016. https://www.darpa.mil/program/common-heterogeneous-integration-and-ip-reuse-strategies

[Gu+16] P. Gu et al., Thermal-aware 3D design for side-channel information leakage, in *Proc. Int. Conf. Comp. Des.* (2016), pp. 520–527. https://doi.org/10.1109/ICCD.2016.7753336

[Gu+18a] P. Gu et al., Cost-efficient 3D integration to hinder reverse engineering during and after manufacturing, in *Proc. Asian Hardw.-Orient. Sec. Trust Symp.* (2018), pp. 74–79. https://doi.org/10.1109/AsianHOST.2018.8607176

[Gu+18b] P. Gu et al., Cost-efficient 3D integration to hinder reverse engineering during and after manufacturing, in *Proc. Asian Hardw.-Orient. Sec. Trust Symp.* (2018), pp. 74–79

[HCM19] A.B. Huang, S.X. Cross, T. Marble, Open source is insufficient to solve trust problems in hardware. How betrusted aims to close the hardware TOCTOU gap. 36C3, minutes 14:35–16:40. Chaos Computer Club (2019). https://media.ccc.de/v/36c3-10690-open_source_is_insufficient_to_solve_trust_problems_in_hardware

[Ime+13b] F. Imeson et al., Securing computer hardware using 3D integrated circuit (IC) technology and split manufacturing for obfuscation, in *Proc. USENIX Sec. Symp.* (2013), pp. 495–510. https://www.usenix.org/conference/usenixsecurity13/technical-sessions/presentation/imeson

[Int19] *Intel Unveils New Tools in Its Advanced Chip Packaging Tool-box*. Intel (2019). https://newsroom.intel.com/news/intel-unveils-new-tools-advanced-chip-packaging-toolbox/

[Iye15] S.S. Iyer, Three-dimensional integration: An industry perspective. MRS Bull. **40**(3), 225–232 (2015). ISSN: 1938-1425. https://doi.org/10.1557/mrs.2015.32

[Jun+17] M. Jung et al., Design methodologies for low-power 3-D ICs with advanced tier partitioning. Trans. VLSI Syst. **25**(7). ISSN: 1063-8210. https://doi.org/10.1109/TVLSI.2017.2670508

[KCL18] B.W. Ku, K. Chang, S.K. Lim, Compact-2D: a physical design methodology to build commercial-quality face-to-face-bonded 3D ICs, in *Proc. Int. Symp. Phys. Des.* Monterey, California (2018), pp. 90–97. ISBN: 978-1-4503-5626-8. https://doi.org/10.1145/3177540.3178244

[Kim+12] D.H. Kim et al., 3D-MAPS: 3D massively parallel processor with stacked memory, in *Proc. Int. Sol.-St. Circ. Conf.* (2012), pp. 188–190. https://doi.org/10.1109/ISSCC.2012.6176969

[Kim+19] J. Kim et al., Architecture, chip, and package co-design flow for 2.5D IC design enabling heterogeneous IP reuse, in *Proc. Des. Autom. Conf.* (2019). https://doi.org/10.1145/3316781.3317775

[Kne18] J. Knechtel. *Proximity Attack for 3D ICs with Obfuscated F2F Mappings*. DfX, NYUAD (2018). https://github.com/DfX-NYUAD/3D-SM-Attack

[KPS19a] J. Knechtel, S. Patnaik, O. Sinanoglu, 3D integration: another dimension toward hardware security, in *Proc. Int. On-Line Test Symp.* (2019), pp. 147–150. https://doi.org/10.1109/IOLTS.2019.8854395

[KS17] J. Knechtel, O. Sinanoglu, On mitigation of side-channel attacks in 3D ICs: decorrelating thermal patterns from power and activity, in *Proc. Des. Autom. Conf.* (2017), pp. 12:1–12:6. https://doi.org/10.1145/3061639.3062293

[Lau11] J.H. Lau, The most cost-effective integrator (TSV Interposer) for 3D IC integration system-in-package (SiP), in *Proc. ASME InterPACK* (2011), pp. 53–63. https://doi.org/10.1115/IPACK2011-52189

[Lee+16] C.C. Lee et al., An overview of the development of a GPU with integrated HBM on silicon interposer, in *Proc. Elec. Compon. Tech. Conf.* (2016), pp. 1439–1444. https://doi.org/10.1109/ECTC.2016.348

[Lee+17] J. Lee et al., Hacking in darkness: return-oriented programming against secure enclaves, in *Proc. USENIX Sec. Symp.* (2017), pp. 523–539. ISBN: 978-1-931971-40-9. https://www.usenix.org/conference/usenixsecurity17/technical-sessions/presentation/lee-jaehyuk

[Li+18a] M. Li et al., A practical split manufacturing framework for trojan prevention via simultaneous wire lifting and cell insertion, in *Proc. Asia South Pac. Des. Autom. Conf.* (2018), pp. 265–270

[LM19] Y. Lyu, P. Mishra, Efficient test generation for Trojan detection using side channel analysis, in *Proc. Des. Autom. Test Europe* (2019), pp. 408–413

[Mae+18] P. Maene et al., Hardware-based trusted computing architectures for isolation and attestation. Trans. Comput. **67**(3), 361–374 (2018). ISSN: 0018-9340. https://doi.org/10.1109/TC.017.2647955

[Mas+15] R.J. Masti et al., Thermal covert channels on multi-core platforms, in *Proc. USENIX Sec. Symp.* (2015), pp. 865–880. ISBN: 978-1-931971-232. https://www.usenix.org/conference/usenixsecurity15/technical-sessions/presentation/masti

[Mav+17] V. Mavroudis et al., A touch of evil: high-assurance cryptographic hardware from untrusted components, in *Proc. Comp. Comm. Sec.* (2017), pp. 1583–1600. https://doi.org/10.1145/3133956.3133961

[McC16] C. McCants, *Trusted Integrated Chips (TIC) Program* (2016). https://www.ndia.org/-/media/sites/ndia/meetings-and-events/divisions/systems-engineering/past-events/trusted-micro/2016-august/mccants-carl.ashx

[MHE17] S.F. Mossa, S.R. Hasan, O. Elkeelany, Self-triggering hardware trojan: Due to NBTI related aging in 3-D ICs. Integration **58**(Supplement C), 116–124 (2017). ISSN: 0167-9260. https://doi.org/10.1016/j.vlsi.2016.12.013

[Mur+20] K. Murdock et al., Plundervolt: software-based fault injection attacks against Intel SGX, in *Proc. Symp. Sec. Priv.* (2020)

[Mys+06] S. Mysore et al., Introspective 3D chips. SIGOPS Oper. Syst. Rev. **40**(5), 264–273 (2006). https://doi.org/10.1145/1168857.1168890

[Nab+20] M. Nabeel et al., 2.5D root of trust: secure system-level integration of untrusted chiplets. Trans. Comput. **69**, 1611–1625 (2020). https://doi.org/10.1109/TC.2020.3020777. Dedicated, after acceptance and publication, in memory of the late Vassos Soteriou; version with dedication note available at https://arxiv.org/abs/2009.02412

[Nab20] M. Nabeel, *HDL framework for 2.5D root of trust*. DfX, NYUAD (2020). https://github.com/DfX-NYUAD/2.5D_ROT

[NG11] *NanGate FreePDK45 Open Cell Library*. Nangate Inc. (2011). http://www.nangate.com/?page_id=2325

[Ngu+08] A. Nguyen-Tuong et al., Security through redundant data diversity, in *Proc. Int. Conf. Depend. Sys. Networks* (2008), pp. 187–196

[Nir+16] I.R. Nirmala et al., A novel threshold voltage defined switch for circuit camouflaging, in *Proc. Europe Test. Symp.* (2016), pp. 1–2. https://doi.org/10.1109/ETS.2016.7519286

[OC19] *Reference community for Free and Open Source IP cores*. Opencores (2019). https://opencores.org/

[OST05] D.A. Osvik, A. Shamir, E. Tromer, Cache attacks and Countermeasures: the Case of AES, in *IACR Crypt. ePrint Arch.* vol. 271 (2005). https://eprint.iacr.org/2005/271

[P D+17] S. M. P. D. et al., A scalable network-on-chip microprocessor with 2.5D integrated memory and accelerator. Trans. Circ. Syst. I **64**(6), 1432–1443 (2017) . ISSN: 1549-8328. https://doi.org/10.1109/TCSI.2016.2647322

[Pat+17] S. Patnaik et al., Obfuscating the interconnects: low-cost and resilient full-chip layout camouflaging, in *Proc. Int. Conf. Comp.-Aided Des.* (2017), pp. 41–48. https://doi.org/10.1109/ICCAD.2017.8203758

[Pat+18c] S. Patnaik et al., Best of both worlds: integration of split manufacturing and camouflaging into a security-driven CAD flow for 3D ICs, in *Proc. Int. Conf. Comp.-Aided Des.* (2018)

[Pat+18e] S. Patnaik et al., Concerted wire lifting: enabling secure and cost-effective split manufacturing, in *Proc. Asia South Pac. Des. Autom. Conf.* (2018), pp. 251–258. https://doi.org/10.1109/ASPDAC.2018.8297314

[Pat+19b] S. Patnaik et al., A modern approach to IP protection and trojan prevention: split manufacturing for 3D ICs and obfuscation of vertical interconnects. Trans. Emerg. Top. Comp. Early Access (2019). https://doi.org/10.1109/TETC.2019.2933572

[Pen+17] Y. Peng et al., Parasitic extraction for heterogeneous face-to-face bonded 3-D ICs. Trans. Compon., Pack., Manuf. Tech. **7**(6), 912–924 (2017). ISSN: 2156-3950. https://doi.org/10.1109/TCPMT.2017.2677963

[PSF17] V.F. Pavlidis, I. Savidis, E.G. Friedman. *Three-dimensional Integrated Circuit Design*, 2nd ed. (Morgan Kaufmann Publishers Inc., 2017). https://www.sciencedirect.com/book/9780124105010/three-dimensional-integrated-circuit-design

[Qiu+19] P. Qiu et al., VoltJockey: Breaching TrustZone by software-controlled voltage manipulation over multi-core frequencies, in *Proc. Comp. Comm. Sec.* (2019), pp. 195–209. ISBN: 978-1-4503-6747-9. https://doi.org/10.1145/3319535.3354201

[Raj+13a] J. Rajendran et al., Security analysis of integrated circuit camouflaging, in *Proc. Comp. Comm. Sec.* Berlin (2013), pp. 709–720. ISBN: 978-1-4503-2477-9. https://doi.org/10.1145/2508859.2516656

[Rod+19] J. Rodriguez et al., LLFI: lateral laser fault injection attack, in *Proc. Worksh. Fault Diag. Tol. Cryptogr.* (2019), pp. 41–47. https://doi.org/10.1109/FDTC.2019.00014

[RSK13] J. Rajendran, O. Sinanoglu, R. Karri, Is split manufacturing secure?, in *Proc. Des. Autom. Test Europe* (2013), pp. 1259–1264. https://doi.org/10.7873/DATE.2013.261

[Sch+19] M. Schwarz et al., ZombieLoad: Cross-privilege-boundary data sampling. Comp. Research Rep. (2019). https://arxiv.org/abs/1905.05726

[Sha+17a] K. Shamsi et al., AppSAT: approximately deobfuscating integrated circuits, in *Proc. Int. Symp. Hardw.-Orient. Sec. Trust* (2017), pp. 95–100. https://doi.org/10.1109/HST.2017.7951805

[Shi18] A. Shilov. *AMD Previews EPYC Rome Processor: Up to 64 Zen 2 Cores* (2018). https://www.anandtech.com/show/13561/amd-previews-epyc-rome-processor-up-to-64-zen-2-cores

[SRM15] P. Subramanyan, S. Ray, S. Malik, Evaluating the security of logic encryption algorithms, in *Proc. Int. Symp. Hardw.-Orient. Sec. Trust* (2015), pp. 137–143. https://doi.org/10.1109/HST.2015.7140252

[ST16] H. Salmani, M.M. Tehranipoor, Vulnerability analysis of a circuit layout to hardware trojan insertion. Trans. Inf. Forens. Sec. **11**(6), 1214–1225 (2016). ISSN: 1556-6013. https://doi.org/10.1109/TIFS.2016.2520910

[Sto+17] D. Stow et al., Cost-effective design of scalable high-performance systems using active and passive interposers, in *Proc. Int. Conf. Comp.-Aided Des.* (2017). https://doi.org/10.1109/ICCAD.2017.8203849

[Synopsys32nm19] *Synopsys 90nm Generic Libraries*. Synopsys (2019). https://www.synopsys. com/community/university-program/teaching-resources.html

[Tak+13] S. Takaya et al., A 100GB/s wide I/O with 4096b TSVs through an active silicon interposer with in-place waveform capturing, in *Proc. Int. Sol.-St. Circ. Conf.* (2013), pp. 434–435. https://doi.org/10.1109/ISSCC.2013.6487803

[Tez08] Tezzaron Semiconductor, *3D-ICs and Integrated Circuit Security*. Tech. rep. Tezzaron Semiconductor (2008). http://tezzaron.com/media/3D-ICs_and_Integrated_Circuit_Security.pdf

[Uhl+18] B. Uhlig et al., Progress on Carbon Nanotube BEOL Interconnects, in *Proc. Des. Autom. Test Europe* (2018)

[Vai+14a] K. Vaidyanathan et al., Building trusted ICs using split fabrication, in *Proc. Int. Symp. Hardw.-Orient. Sec. Trust* (2014), pp. 1–6. https://doi.org/10.1109/ HST.2014.6855559

[Val+13] J. Valamehr et al., A 3-D split manufacturing approach to trustworthy system development. Trans. Comput. Aided Des. Integr. Circ. Syst. **32**(4), 611–615 (2013). ISSN: 0278-0070. https://doi.org/10.1109/TCAD.2012.2227257

[Viv+20] P. Vivet et al., A 220GOPS 96-core processor with 6 chiplets 3D-stacked on an active interposer offering 0.6ns/mm Latency,3Tb/s/mm2 inter-chiplet interconnects and 156mW/mm2@ 82%-peak-efficiency DC-DC converters, in *Proc. Int. Sol.-St. Circ. Conf.* (2020), pp. 46–48. https://doi.org/10.1109/ ISSCC19947.2020.9062927

[Wah+16] R.S. Wahby et al., Verifiable ASICs, in *Proc. Symp. Sec. Priv.* (2016), pp. 759–778. https://doi.org/10.1109/SP.2016.51

[Wan+15a] C. Wang et al., TSV-based PUF circuit for 3DIC sensor nodes in IoT applications, in *Proc. Electron. Dev. Solid State Circ.* (2015), pp. 313–316. https://doi.org/10.1109/EDSSC.2015.7285113

[Wan+18b] Y. Wang et al., The cat and mouse in split manufacturing. Trans. VLSI Syst. **26**(5), 805–817 (2018). ISSN: 1063-8210. https://doi.org/10.1109/TVLSI. 2017.2787754

[WYM14] M. Wang, A. Yates, I.L. Markov, Super-PUF: integrating heterogeneous physically unclonable functions, in *Proc. Int. Conf. Comp.-Aided Des.* San Jose, California (2014), pp. 454–461. ISBN: 978-1-4799-6277-8. https://doi. org/10.1109/ICCAD.2014.7001391

[XBS17] Y. Xie, C. Bao, A. Srivastava, Security-aware 2.5D integrated circuit design flow against hardware IP piracy. Computer **50**(5), 62–71 (2017). ISSN: 0018-9162. https://doi.org/10.1109/MC.2017.121

[Yan+18] C. Yan et al., Hardware-efficient logic camouflaging for monolithic 3D ICs. Trans. Circ. Syst. II **65**(6), 799–803 (2018). ISSN: 1549-7747. https://doi.org/ 10.1109/TCSII.2017.2749523

[Yin+18] J. Yin et al., Modular routing design for chiplet-based systems, in *Proc. Int. Symp. Comp. Archit.* (2018), pp. 726–738. https://doi.org/10.1109/ISCA. 2018.00066

[Yu+17] C. Yu et al., Incremental SAT-based reverse engineering of camouflaged logic circuits. Trans. Comput. Aided Des. Integr. Circ. Syst. **36**(10), 1647–1659 (2017). ISSN: 0278-0070. https://doi.org/10.1109/TCAD.2017.2652220

Chapter 7
Tamper-Proof Hardware from Emerging Technologies

7.1 Chapter Introduction

Physical attacks and tampering on Integrated Circuits (IC) have become a menace in today's interconnected world, where there are various avenues of infringing on protected hardware, ranging from temperature and magnetic probe-based attacks [Roh09, Gho16] to focused ion beam (FIB) [TJ09] and laser probing from the chip backside [Kin14, Qua+16a]. The attacker in this case could have a multitude of end goals, including but not limited to (i) stealing confidential data, (ii) reverse engineering proprietary design intellectual property (IP), or simply (iii) causing a denial-of-service (DoS).

As attackers up their game by progressing to advance techniques like optical probing based on photo-emission and electro-optical frequency modulation, laser voltage-based detection [Wan+17b], plasma FIB delayering [Pri+17b], and electromagnetic near-field coupling probes [Rao+15], the design house is forced to invest in and employ novel security schemes to circumvent these malicious incursions. Emerging devices and materials, by the virtue of their tamper-resilient properties, might be able to tip the scales of this equation in favor of the designer.

In this chapter, we explore some of these cutting-edge material systems and emerging architectures and look at what intrinsic properties make them good candidates for implementing tamper-proof and secure electronics. First, we establish the key concepts and ideas for developing tamper-proof hardware from emerging devices, followed by a review of the seminal works that have successfully achieved the same. Finally, we examine the case study of a promising secure memory solution in detail.

7.2   Concepts for Tamper-Proof Logic and Memory

In order to forge an emerging technology-based tamper-proofing primitive, first, the appropriate material system or device has to be selected. This can be done once the security requirements of the system are known, in terms of the expected attack scenarios against which the system is to be protected. Typically, the critical vulnerabilities in logic and memory systems, with respect to physical tampering-based attacks, concern IP protection and data protection, respectively. The common threat to both arises in the form of DoS attacks. Although there are various approaches to tackle these attack avenues, there is rarely a catch-all solution to secure the target system from all malicious elements:

1. The most commonly employed scheme for tamper-proofing is the **sensor-based approach**. The target system here is fitted with sensor devices, capable of monitoring any changes in the physical parameters in their vicinity. These can include changes in the temperature, resistance, capacitance, magnetic field, etc. Any tampering attack on the system is highly likely to leave a footprint and change these parameters, thus triggering an alarm. Once the incursion is identified, the system can then decide to shut down temporarily/permanently or flush sensitive data and keys. Emerging device-based sensor configurations are promising for such applications due to advantages like extremely high sensitivity, low power operation, and small area.
2. Emerging technologies can also achieve tamper-proofing by virtue of their **novel architecture and circuit design**. For instance, 2.5D and 3D vertical integration can enable the placement of critical hardware in the middle of the stack, wherein the complexity for an attack from the top and bottom is increased. Additional wires and mesh structures can be placed to interfere with any probing attempt [KPS19b].
3. The security characteristics imparted by **novel materials** can be exploited to augment tamper-proof systems. In particular, certain materials like antiferromagnets and superconductors are known to be impervious to [Ran+20b] or expel magnetic fields [Hir12], respectively. This can be leveraged to design memory systems capable of thwarting electromagnetic probing attacks. Note however that, taking implementation cost into account, the cryogenic cooling required to maintain superconductivity might prohibit the use of such materials.
4. Conventionally, **cryptographic integrity verification** using various cipher modules has been predominantly used with CMOS [Mac20] systems. Although such security solutions do not necessarily utilize emerging technologies, we mention them here for the sake of completeness.

7.3   Review of Selected Emerging Technologies and Prior Art

In this section, we briefly review prior emerging device-based tamper-proof hardware implementations. The purpose of this survey is to give an insight into how emerging devices and materials can help protect our computing and memory systems from physical attacks, in various settings.

7.3.1   Read-Proof Hardware from Protective Coatings

Pim Tuyls et al. introduced in [Tuy+06] one of the first physical implementations of read-proof hardware, resistant against invasive attacks seeking to modify the device structure in order to gain access to internal data. They identified the vulnerability arising out of the practice of permanently storing the digital key in a memory register in the IC. To circumvent this threat, they proposed constructing the key at runtime using an on-chip physically unclonable function (PUF). The premise here is that, since the PUF's challenge-response behavior characterizes its structure completely, any physical tampering or damage in the PUF layer would be evidenced by a significant change in its challenge-response behavior. This implementation of a read-proof storage device is particularly useful for securing smart cards, trusted platform modules, RFID tags, etc., and also for storing the secret key in logic locking schemes and in cryptographic cores like AES.

The read-proof hardware in [Tuy+06] is constructed using a Coating PUF, which is (i) opaque to destructive and non-destructive measurements, (ii) physically and mathematically unclonable, and (iii) tamper-evident. A Coating PUF, shown in Fig. 7.1, consists of a layer of protective coating around the IC. This protective layer is composed of aluminophosphate matrix material doped with a mixture of TiN and TiO_2 dielectric particles of random shapes, sizes, and locations. These dopants impart the PUF layer with opaqueness, hardness, and conductivity. The matrix material makes the coating chemically inert.

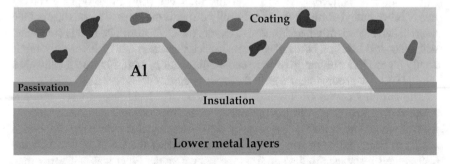

Fig. 7.1 Cross-section of a Coating PUF deposited on top of an IC. The top metal layer contains Al-based sensors, which measure the local capacitance of the coating

The working of the read-proof hardware is as follows. The top metal layer of the IC beneath the PUF coating is fitted with multiple Al-based capacitance sensors, capable of measuring local capacitance values of the PUF. A reference capacitance measurement, obtained under normal operating conditions, is first used to reconstruct the unique fingerprint or key (K) of the Coating PUF through a helper data algorithm/fuzzy extractor. After the PUF capacitance is characterized and the IC is deployed in the field, any physical incursion in the form of drilling, milling, FIB, etc., results in deviations in the PUF dielectric capacitance. This deviation is picked up by the on-chip sensors, which produce a difference fingerprint (K') this time. The discrepancy in the PUF challenge-response behavior during the attempted attack makes any recovered key futile.

Since the sensing and measurement setups are under the coating layer, they cannot be modified by the attacker without triggering a capacitance change. The opaqueness of the coating prevents any inspection of the PUF response from the outside. Any non-contact measurement of the capacitance from the outside is ineffective since the measurements are anisotropic with respect to the locations of the dielectric particles in the coating. Overall, this read-proof scheme is easy and cheap to integrate on-chip and offers good robustness against physical, chemical, and removal attacks.

7.3.2 Carbon Nanotube-Based Sensors for Cryptographic Applications

Rather than the capacitance sensor-based approach in Sect. 7.3.1, Boday et al. introduced a carbon nanotube (CNT) resistance sensor in [Bod+14]. Such a CNT sensor array can be implanted in a wide range of security-critical components and environments, to make logic and memory systems impervious to tampering and physical attacks. In fact, their technique has been demonstrated on a security card to achieve Level-4 Federal Information Processing Standards (FIPS) protection.

The sensor implementation process in [Bod+14] involves, firstly, the CNT film formation. This is accomplished by mixing the CNT rings in an organic solvent like N,N'-dimethyl-formamide, to uniformly disperse the rings. The mixture is then filtered, set, heated, dried, and peeled to extract the CNT film. The film is connected with electrical leads on both sides, for resistance measurement. This sensor setup is finally placed in a polymeric resin to encase the security card. Figure 7.2 illustrates the CNT sensor arrangement within the security card.

The working of the CNT sensor is as follows. Each fabricated CNT film has a known fixed resistance in the unstrained state. This original resistance value is first measured and stored on the security card. Any physical tampering of the card results in compressive or tensile strain, which stretches the CNTs in the resin and causes their resistance to deviate from the known value. Electrical tampering may cause a short in the CNT array and hence again change the film resistance. Any

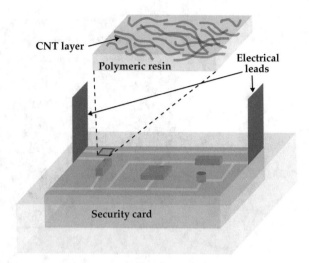

Fig. 7.2 CNT sensor configuration within security card. The CNTs are encased in a polymeric resin and connected to electrical leads on the card, to measure resistance changes

resistance variation outside the nominal operating range of the card is detected by the measurement circuitry, which can then trigger protection mechanisms meant to shut down card functionality or erase its contents. Moreover, temperature-based attacks on the card will change the physical state of the polymeric encapsulation, also resulting in a CNT resistance change. Hence, attackers aiming to leverage liquid nitrogen or heat to delaminate the polymeric encapsulation layer are thwarted.

Securing cryptographic systems with such polymer encased CNT sensors have the added advantage of providing enhanced structural strength from the CNT fillers, while implementing an extremely sensitive resistance sensor at the macro level.

7.3.3   3D Hardware Cage

3D integration and vertical silicon technology enables novel security and tamper-proofing solutions. For instance, 3D Hamiltonian paths can be used as hardware integrity verification sensors in 3D ICs [Bri+12b]. The Hamiltonian path, which is the 3D cage here, is an undirected path that visits all the vertices of a graph, exactly once. Briais et al. propose the concept of a hardware canary in [Bri+12b], which is used as a sensor and alarm system to thwart tampering in their 3D cage. It consists of a chain of cryptographic function blocks f_i, placed at the vertices of the Hamiltonian graph, to construct the 3D cage. Given a challenge C to the 3D canary circuit, only the correct response, corresponding to $(f_n \circ \ldots \circ f_i)(C)$, can authenticate the canary's path integrity.

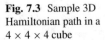

Fig. 7.3 Sample 3D
Hamiltonian path in a
$4 \times 4 \times 4$ cube

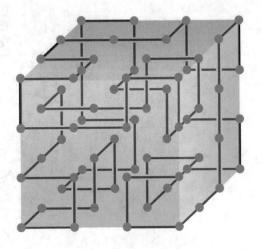

The vertical stacking in 3D integration allows one to design Hamiltonian paths, which go around the entire 3D IC and form a cage to protect the chip from physical probing and tampering attacks. Briais et al. introduce algorithms for constructing random 3D Hamiltonian cycles that go through the internal transistors of the protected IC. Such a path routed through the different metal layers and vertical stacks of the 3D IC can realize a digital integrity verification sensor, by means of sending challenge signals and comparing the responses. Further, the algorithm can be programmed to synthesize Hamiltonian paths, wherein the wires of the path can cover the empty spaces inside the internal circuit. This reinforces the compactness of the 3D IC and increases reverse engineering complexity. Interested readers are referred to [Bri+12b] for further details on the Hamiltonian path-generating algorithms. A sample Hamiltonian path for a $4 \times 4 \times 4$ cube is shown in Fig. 7.3.

The vertices of the Hamiltonian path are implemented with a 3D switch box, comprising the following: (i) a message authentication code (MAC) block realizing the cryptographic function f_i, (ii) a register for storing the result of the MAC, (iii) two multiplexers for properly routing the I/O, and (iv) a control block for configuring the multiplexers. If an attacker attempts to tamper with any of the switch boxes in the path, then the final response $(f_n \circ \ldots \circ f_i)(C)$ is altered. This deviation is detected by XORing the modified response with the response from a mirror verification circuit. Hence, the output of the XOR determines if there has been a tampering attack on the IC. Under attack conditions, the safety protocol of the IC is initiated and the sensitive data in the registers is erased. The 3D cage serves the purpose of physically protecting the XOR circuit as well, preventing any removal attacks.

7.4 Case Study: Secure Magnetoelectric Antiferromagnet-Based Tamper-Proof Memory

Conventional dynamic random access memory (DRAM) scaling has reached a critical tipping point as the miniaturization of the DRAM cell has plateaued in recent years. A promising solution to the memory scaling problem is to realize the main memory system using non-volatile technologies [Mut15]. While NVMs offer attractive features, such as high density, low leakage, and non-volatile data retention, they also suffer from poor endurance and high access latency in their current implementation.

Memory security has come under more scrutiny over the years. This is because of attacks such as *Spectre* [Koc+19b] and *Meltdown* [Lip+18b], which target the side-channels associated with speculative execution and out-of-order execution, respectively, have exposed severe vulnerabilities in a wide array of currently deployed processors and their memory architectures. In the case of NVMs, data remanence after power-down presents a severe threat to data confidentiality, as attackers aiming to steal private data can do so easily by mounting cold-boot attacks [Hal+09] or other removal attacks like stealing the memory module (DIMM) [YNQ15]. Moreover, magnetic memories like the spin-transfer torque magnetoresistive random access memory (STT-MRAM) are highly sensitive to stray magnetic fields. As such, magnetic field-based attacks [Jan+15] can be used to corrupt the stored data or compromise the memory's functional integrity, resulting in a DoS attack. Hence, such security vulnerabilities pose a significant impediment to the pervasive and large-scale proliferation of NVMs in the memory industry.

Prior works on securing NVMs have focused mainly on memory encryption schemes, which are necessary to prevent attackers from exploiting data remanence in the off-state. Chhabra et al. proposed an incremental encryption scheme [CS11] for NVMs where only inert memory pages, which have not been accessed for several clock cycles, are encrypted selectively. A sneak-path encryption (SPE) scheme was demonstrated for memristor-based NVMs in [KKS14], wherein sneak paths in the memristor crossbar array are exploited to apply encryption pulses to change the resistances of the memory cells and hence encrypt the stored data. In [YNQ15], the authors proposed DEUCE, a dual counter encryption for phase change memories (PCM), which significantly reduces the number of modified bits per writeback, to improve performance and lifetime of the memory. Swami et al. took this concept forward and proposed SECRET [SRM16], a smart encryption scheme for NVMs, which integrates word-level re-encryption and zero-based partial writes to reduce memory write operations. An advanced counter mode encryption (ACME) was presented in [SM18], which utilizes the write leveling architecture inherent in PCM memories, to perform counter-write leveling. ACME helps to avoid *Rowhammer*-type attacks by preventing the counter associated with any single cache line from overflowing. The impact of contactless tampering on STT-MRAMs using external magnetic fields was highlighted in [Jan+15]. Using micromagnetic simulations, the authors of [Jan+15] showed how magnetic field-based attacks could corrupt the contents of STT-MRAM cells. Techniques to protect against contactless attacks proposed in [Jan+15] included (i) an on-chip sensor to detect magnetic field-based incursions and (ii) error correction modules to compensate cell failures arising due to magnetic field attacks. However, these techniques incur large energy and area penalties due to the additional hardware imposed by the magnetic field sensor and the error correction scheme.

In [Ran+20b], an alternative to conventional NVMs is presented in the form of *SMART: A Secure Magnetoelectric Antiferromagnet-Based Tamper-Proof Non-Volatile Memory*. SMART memory leverages the room-temperature linear magnetoelectric (ME) effect in antiferromagnets (AFMs) like chromia [RF], which can be switched solely using voltage pulses, without the use of electric currents. In terms of secure and tamper-proof data storage, SMART memory offers a significant advancement as compared to the existing memory systems. For example, AFMs do not exhibit a magnetic signature since they do not have a net external magnetic moment, unlike ferromagnets (FM). Hence, the SMART memory cannot be probed or switched with external magnetic fields, unlike the way STT-MRAMs can. This, in turn, eliminates the possibility of magnetic field attacks undermining data integrity or aiming to induce DoS. To address the post-shutdown data remanence of SMART memory, an in-memory encryption scheme employing controlled-NOT (CNOT) logic is implemented. SMART memory addresses attacks aiming to compromise data confidentiality and data fidelity, in both powered-on and powered-off states.

7.4.1   Construction and Working of SMART

The chromia-based magnetoelectric antiferromagnetic random access memory (ME-AFMRAM), which is at the heart of the SMART memory, is shown in Fig. 7.4. Experimentally demonstrated by Kosub et al. [Kos+17], the ME-AFMRAM has a

Fig. 7.4 Chromia-based magnetoelectric antiferromagnetic random access memory. Data (1/0) is written by applying a voltage (+/−) to the bottom gate electrode. Read-out is achieved using an anomalous Hall bar electrode placed on top, by applying a Hall bias

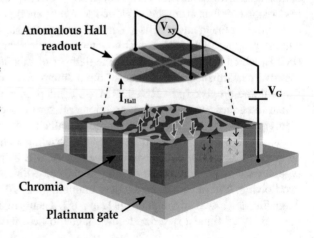

bottom gate electrode (Platinum gate in the figure) for applying the gate voltage V_G and providing the necessary electric field to write data into the memory. The magnetoelectric (ME) effect in antiferromagnetic chromia provides an energy-efficient, all-electrical switching of the roughness-insensitive surface magnetization [EB13]. This surface magnetization is strongly coupled to the AFM order parameter. That is, the electrical switching of the AFM order results in reversal of the surface magnetization [Wu+11], which is used to encode the information in the ME-AFMRAM.

A small, symmetry-breaking magnetic field (≈ 30 mT) is provided by the stray field of a permanent magnet. A positive voltage V_G will orient the bulk order and, hence, put the surface magnetization in one domain (with surface moments pointing up), whereas a negative voltage will result in the surface magnetization relaxing to the opposite domain (with surface moments pointing down). These two states correspond to binary levels "1" ($V_G > 0$) and "0" ($V_G < 0$), respectively. A gate voltage of 0 V corresponds to the "hold" mode of the memory cell. The read-out is achieved using an anomalous Hall (AH) bar electrode setup, which discerns the surface magnetization of chromia by sensing the proximity effect-induced magnetization in the nearby Platinum (Pt) electrode, thereby producing a proportional Hall voltage V_{xy} (or V_{AHE}) [Kos+15].

The individual ME-AFRAM cells are grouped into sub-arrays along with the peripheral decoders, drivers, and sense amplifiers, as highlighted in Fig. 7.5. These sub-arrays are further organized into larger memory systems as shown in Fig. 7.6 [Don+12].

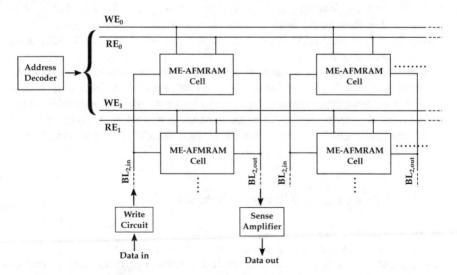

Fig. 7.5 Construction of the ME-AFMRAM cell array. The signals $BL_{i,in}$ serve to write data into the cells when Write Enable (WE) is on, and signals $BL_{i,out}$ serve to read data from the cells when Read Enable (RE) is on

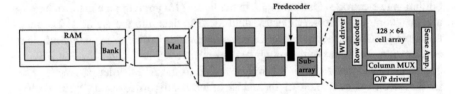

Fig. 7.6 64 KB ME-AFMRAM organization with 4 × 1 banks, 2 × 1 mats, 4 × 2 sub-arrays, and 128 × 64 bit cell arrays. Here, the word length is 128 bit. The memory organization is leveraged from [Don+12]

7.4.2 Application of SMART for Secure Storage

7.4.2.1 Threat Model

The threat model, which defines the strengths and capabilities of attackers as well as the objectives and consequences of a successful attack on SMART, is described next:

- Attackers can launch cold-boot attacks [Hal+09]. During power-down, there is some latency after the power-down sequence initiates until the moment when memory contents are completely secured. An attacker might use this gap to read-out memory contents. To circumvent such attacks, memory encryption is typically employed [CS11, SM18].
- Attackers could leverage properties like sensitivity to magnetic fields and temperature fluctuations to corrupt the data or induce a DoS [Jan+15]. They may forcibly write specific data patterns to memory, which accelerates aging and causes memory failures.
- With access to failure analysis equipment, attackers can also resort to advanced invasive attacks. The majority of such attacks target at the back-end-of-line (BEOL), approaching from the top-most metal layer, which is also referred to as front-side attacks. Various countermeasures have been proposed to protect the front-side, which include protective meshes, shields, and sensors [Lee+19, Wei+18].

7.4.2.2 Magnetic Field and Temperature Attacks

STT-MRAMs have FM-based magnetic tunnel junctions (MTJ) as their basic building blocks. FMs possess a macroscopic magnetization (or magnetic signature) that can be probed or inferred with using an external magnetic field. Hence, magnetic fields can be used to infer or tamper with the stored data or even cause malfunctions in STT-MRAMs [Jan+15]. Stray magnetic fields as small as 10 mT could cause an unintended bit flip in STT-MRAM cells. Figure 7.7 shows the

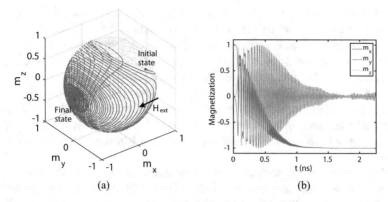

Fig. 7.7 The FMs in an STT-MRAM can be switched easily using external magnetic fields. (**a**) Trajectory for magnetic field-induced switching of a FM. (**b**) Components for magnetic field-induced switching of a FM

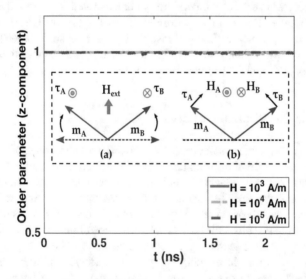

Fig. 7.8 The application of a magnetic field is unable to switch the AFM order parameter, even when increasing the field magnitude. Inset: (**a**) an external, homogeneous magnetic field may cant the sublattice moments, but it is incapable of rotating the AFM order; (**b**) staggering fields on the sublattice moments produce staggered tangential torques, which can reorient the AFM order

magnetic field-induced bit flip in a representative FM, obtained by solving the Landau–Lifshitz–Gilbert equation for the FM dynamics [Ame+16].

AFMs, on the other hand, exhibit no external magnetic signature since their equal and opposite sublattice moments cancel each other out. Hence, the bulk order parameter cannot be affected by external magnetic fields. To switch the bulk order, staggered fields (opposite sign on opposite sublattices) must be applied on both the sublattice moments, as illustrated in Fig. 7.8 inset. However, an external, homogeneous magnetic field is unable to provide such a staggered field arrangement and, hence, ends up canting the sublattice moments in a way wherein the torque due to the external field is exactly balanced by the exchange torque exerted by one sublattice moment on the other [Bal+18]. Since external magnetic fields are unable to reorient the AFM order parameter, the SMART ME-AFMRAM is expected to be resistant to magnetic field attacks described in [Jan+15]. Note that switching the ME-AFM surface magnetization state using a combination of E and H fields would require an exact knowledge of the write cycles and the prior state of the surface, as well as means to control the electric field explicitly, which is to be concealed from an attacker.

With regard to temperature fluctuation-based attacks, an adversary might increase the ambient temperature of the ME-AFMRAM in an attempt to alter the stored data. Note that the Néel temperature of pure chromia is 308 K [SWB09], above which the AFM ordering is destroyed. Hence, the attacker may corrupt the memory by heating it above the Néel temperature. To counter this, Boron-doped chromia, whose Néel temperature is demonstrated to be ~400 K [Str+14], is utilized. Hence, Boron-doped chromia increases the resilience of SMART memory against temperature fluctuations. That is because such larger temperature fluctuations (above 400 K) are easier to detect, and countermeasures like interception of such attacks become more feasible.

7.4.2.3   Data Confidentiality Attacks

As with all NVMs, data remanence in the SMART memory could be exploited by attackers to steal sensitive information. The most effective countermeasure against such data confidentiality attacks, including cold-boot and stolen memory modules attacks, is to encrypt the data using a secure encryption scheme before storing it in the memory. Advanced memory encryption techniques like counter mode encryption (CME) use block ciphers such as Advanced Encryption Standard (AES) to encrypt a seed using a secret key, in order to generate a one-time pad (OTP).

SMART memory uses in-memory encryption, or *Memcryption*, instead of XOR-based CME schemes, to optimize the overall memory access latency and avoid idle clock cycles. *Memcryption* uses bitwise CNOT gates constructed from ME-AFM-based logic, wherein the encryption pulse is tied to the control signals of CNOT gates. Spin devices like the ME-AFM transistor [Dow+18], which are able to implement polymorphic logic gates, can be used to realize the CNOT gate. In *Memcryption*, ME-AFM transistor-based CNOT gates are embedded directly in the data path connected to the memory array; hence, the encryption is in-memory, as opposed to prior works using a separate encryption block.

The SMART memory architecture with *Memcryption* is shown in Fig. 7.9. A trusted 128-bit key, provided and stored within a secure processing module (SPM) along with the processor, is concatenated with the memory address and used as seed for AES. The AES core, which is to be integrated on the NVM chip, thus produces an encryption pulse whose bits are used as the control bits for the

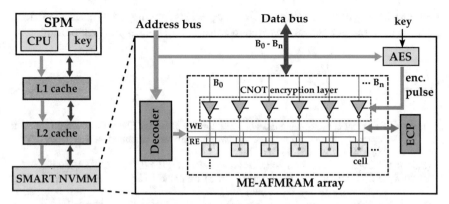

Fig. 7.9 SMART memory architecture with *Memcryption*. The CNOT layer for decryption is not shown for simplicity

CNOT gates of the in-memory encryption layer. Depending on the control bits, the encryption layer flips bits selectively in the plaintext before performing a memory write. During decryption, the same encryption pulse is generated again and used to perform bitwise CNOT operations on the cipher-text (read from memory), to obtain the plaintext. The architecture also includes an error correction pointer (ECP) scheme [SM17] to address faults in the CMOS peripherals and ensure the correctness of the stored data.

7.5 Closing Remarks

The growing threat of physical attacks has necessitated the materials research-driven design of new secure computing and storage systems. The unique electrical, physical, and mechanical attributes of novel materials and phenomena have enabled a bottom-up modular approach in the device-circuit co-design of modern security primitives, wherein a particular material can be mapped to the required level of security and resilience to attacks. The same can be said about novel architectural design processes that promote security by construction. For instance, CNT-based sensors can pick up changes in resistance due to electrical/mechanical tampering, antiferromagnetic devices can be made impervious to external magnetic fields, and 3D integration allows unique spatial authentication techniques based on Hamiltonian paths.

This chapter delved into how such material and architectural innovations in the emerging technology space can open up new avenues for protecting our logic and memory systems from physical incursions. First, the primary concepts concerning tamper-resilience were delineated, followed by a review of the seminal emerging technology-based tamper-proof hardware in the literature. Then a case study on a secure memory solution, using antiferromagnets, is highlighted to give the reader a sense of the design practices involved in constructing tamper-resilient hardware.

References

[Ame+16] S. Ament et al., Solving the stochastic Landau-Lifshitz-Gilbert-Slonczewski equation for monodomain nanomagnets: a survey and analysis of numerical techniques (2016). Preprint. arXiv:1607.04596

[Bal+18] V. Baltz et al., Antiferromagnetic spintronics. Rev. Mod. Phys. **90**(1), 015005 (2018)

[Bod+14] D.J. Boday et al., Implementing carbon nanotube based sensors for cryptographic applications (2014). US Patent 8,797,059

[Bri+12b] S. Briais et al., 3D hardware canaries, in *International Workshop on Cryptographic Hardware and Embedded Systems* (Springer, Berlin, 2012), pp. 1–22

[CS11] S. Chhabra, Y. Solihin, i-NVMM: a secure non-volatile main memory system with incremental encryption, in *Computer Architecture (ISCA), 2011 38th Annual International Symposium on* (IEEE, Piscataway, 2011), pp. 177–188

[Don+12] X. Dong et al., NVSim: a circuit-level performance, energy, and area model for emerging nonvolatile memory. IEEE Trans. Comput. Aided Des. Integr. Circuits Syst. **31**(7), 994–1007 (2012)

[Dow+18] P.A. Dowben et al., Towards a strong spin-orbit coupling magneto-electric transistor. IEEE J. Explor. Solid-State Comput. Dev. Circ. **4**(1), 1–9 (2018)

[EB13] W. Echtenkamp, Ch. Binek, Electric control of exchange bias training. Phys. Rev. Lett. **111**(18), 187204 (2013)

[Gho16] S. Ghosh, Spintronics and security: prospects, vulner-abilities, attack models, and preventions. Proc. IEEE **104**(10), 1864–1893 (2016)

[Hal+09] J.A. Halderman et al., Lest we remember: cold-boot attacks on encryption keys. Commun. ACM **52**(5), 91–98 (2009)

[Hir12] J.E. Hirsch, The origin of the Meissner effect in new and old superconductors. Physica Scripta **85**(3), 035704 (2012)

[Jan+15] J.-W. Jang et al., Self-correcting STTRAM under magnetic field attacks, in *2015 52nd ACM/EDAC/IEEE Design Automation Conference (DAC)* (IEEE, Piscataway, 2015), pp. 1–6

[Kin14] U. Kindereit, Fundamentals and future applications of laser voltage probing, in *2014 IEEE International Reliability Physics Symposium* (IEEE, Piscataway, 2014), 3F–1

[KKS14] S. Kannan, N. Karimi, O. Sinanoglu, Secure memristor-based main memory, in *2014 51st ACM/EDAC/IEEE Design Automation Conference (DAC)* (IEEE, Piscataway, 2014), pp. 1–6

[Koc+19b] P. Kocher et al., Spectre attacks: exploiting speculative execution, in *2019 IEEE Symposium on Security and Privacy (SP)* (IEEE, Piscataway, 2019), pp. 1–19

[Kos+15] T. Kosub et al., All-electric access to the magnetic-field-invariant magnetization of antiferromagnets. Phys. Rev. Lett. **115**(9), 097201 (2015)

[Kos+17] T. Kosub et al., Purely antiferromagnetic magnetoelectric random access memory. Nat. Commun. **8**, 13985 (2017)

[KPS19b] J. Knechtel, S. Patnaik, O. Sinanoglu, 3D integration: another dimension toward hardware security, in *2019 IEEE 25th International Symposium on On-Line Testing and Robust System Design (IOLTS)* (IEEE, Piscataway, 2019), pp. 147–150

[Lee+19] Y. Lee et al., Robust secure shield architecture for detection and protection against invasive attacks. Trans. Comp. Aided Des. Integ. Circ. Sys. (2019). ISSN: 1937-4151. https://doi.org/10.1109/TCAD.2019.2944580

[Lip+18b] M. Lipp et al., Meltdown: reading kernel memory from user space, in *27th {USENIX}Security Symposium ({USENIX}Security 18)* (2018), pp. 973–990

[Mac20] Private Machines. Next-Gen IT Infrastructure Protection (2020). https://privatemachines.com/briefs/Private.Machines.ENFORCER.Overview.1page.pdf

[Mut15] O. Mutlu, Main memory scaling: challenges and solution directions, in *More than Moore Technologies for Next Generation Computer Design* (Springer, Berlin, 2015), pp. 127–153

[Pri+17b] E.L. Principe et al., Plasma FIB deprocessing of integrated circuits from the backside, in *FICS Research Annual Conference on Cybersecurity* (2017)

[Qua+16a] S.E. Quadir et al., A survey on chip to system reverse engineering. ACM J. Emerg. Technol. Comput. Syst. **13**(1), 1–34 (2016)

[Ran+20b] N. Rangarajan et al., SMART: a secure magnetoelectric antiferromagnet-based tamper-proof non-volatile memory. IEEE Acces. **8**, 76130–76142 (2020)

[Rao+15] J. Raoult et al., Electromagnetic coupling circuit model of a magnetic near-field probe to a microstrip line. In: *2015 10th International Workshop on the Electromagnetic Compatibility of Integrated Circuits (EMC Compo)* (IEEE, Piscataway, 2015), pp. 29–33

[RF] G.T. Rado, V.J. Folen, Observation of the magnetically induced magnetoelectric effect and evidence for antiferromagnetic domains. Phys. Rev. Lett. **7**(8), 310

[Roh09] P. Rohatgi, Electromagnetic attacks and countermeasures, in *Cryptographic Engineering* (Springer, Berlin, 2009), pp. 407–430

[SM17] S. Swami, K. Mohanram, Reliable nonvolatile memories: techniques and measures. IEEE Desig. Test **34**(3), 31–41 (2017)

[SM18] S. Swami, K. Mohanram, ACME: advanced counter mode encryption for secure non-volatile memories. in *2018 55th ACM/ESDA/IEEE Design Automation Conference (DAC)* (IEEE, Piscataway, 2018), pp. 1–6

[SRM16] S. Swami, J. Rakshit, K. Mohanram, SECRET: smartly encrypted energy efficient non-volatile memories, in *Proceedings of the 53rd Annual Design Automation Conference* (2016), pp. 1–6

[Str+14] M. Street et al., Increasing the Néel temperature of magneto-electric chromia for voltage-controlled spintronics. Appl. Phys. Lett. **104**(22), 222402 (2014)

[SWB09] S. Shi, A.L. Wysocki, K.D. Belashchenko, Magnetism of chromia from first-principles calculations. Phys. Rev. B **79**(10), 104404 (2009)

[TJ09] R. Torrance, D. James, The state-of-the-art in IC reverse engineering, in *International Workshop on Cryptographic Hardware and Embedded Systems* (Springer, Berlin, 2009), pp. 363–381

[Tuy+06] P. Tuyls et al., Read-proof hardware from protective coatings, in *International Workshop on Cryptographic Hardware and Embedded Systems* (Springer, Berlin, 2006), pp. 369–383

[Wan+17b] H. Wang et al., Probing attacks on integrated circuits: challenges and research opportunities. IEEE Desig. Test **34**(5), 63–71 (2017)

[Wei+18] M. Weiner et al., The low area probing detector as a countermeasure against invasive attacks. Trans. VLSI Syst. **26**(2), 392–403 (2018). ISSN: 1063-8210. https://doi.org/10.1109/TVLSI.2017.2762630

[Wu+11] N. Wu et al., Imaging and control of surface magnetization domains in a magnetoelectric antiferromagnet. Phys. Rev. Lett. **106**(8), 087202 (2011)

[YNQ15] V. Young, P.J. Nair, M.K. Qureshi, DEUCE: write-efficient encryption for non-volatile memories. ACM SIGARCH Comput. Archit. News **43**(1), 33–44 (2015)

Chapter 8
Resilience Against Side-Channel Attacks in Emerging Technologies

8.1 Chapter Introduction

It is widely recognized among the security community that devising hardware implementations of cryptographic algorithms and other sensitive computation remains a fundamental challenge. Among various threats arising for real-world application of security-critical electronics, an adversary can observe and measure physical effects ranging from timing, power, electromagnetic (EM) radiation, photon emission, etc.—such side-channels and related attacks (SCAs) are effective for information leakage, relatively easily to exploit, and difficult to fully defend against [ZF05]. For example, more than two decades have passed since the inception of the seminal power side-channel (PSC) attack proposed by Kocher et al. [KJJ99], but PSC attacks remain a significant threat to the security of electronic systems, e.g., see [SW12, OD19].

This chapter investigates the various device- and circuit-level properties from which side-channel resilience and vulnerabilities can arise. It provides an overview of several emerging technologies and related prior art that can accordingly mitigate PSC attacks. Finally, two case studies are presented; the first studies the prospects of power side-channel attacks on negative capacitance field effect transistors (NCFETs), whereas the second details the impact of thermal side-channel attacks on 3D ICs. Both studies investigate the specific root-causes for the side-channel attacks and the roles the respective technologies play. Both studies also cover promising design-time measures to hinder the related attacks.

8.2 Concepts for Side-Channel Resilience with Emerging Technology-Based Circuits

The task of determining the side-channel resilience of any technology involves studying the intrinsic physical and operational properties in the basic logic and memory elements of that technology, as these are key to either mitigate or advance side-channel leakage. In the case of CMOS, side-channel leakage arises due to a variety of factors, like asymmetric and input-dependent power consumption during logic transitions or EM radiation and photon emission during transistor switching. Such security-critical properties require complex countermeasures like masking of the intermediate logic transitions, or hiding the power consumption among additionally induced noise within the circuit. However, emerging technologies may inherently provide defense measures, owing to their unique properties. Some examples are outlined next.

1. Emerging technologies differ from conventional CMOS technologies in terms of their underlying working principles, dynamics of operation, and the physical phenomena driving them. This means that the devices may not share the same side-channel vulnerabilities that plague CMOS, including power, photonic, and EM side-channels. For instance, spintronic devices that do not operate on the principle of carrier generation and recombination do not emit any photons and, hence, are immune to photonic side-channel attacks [Pat+18a]. Careful consideration must be given to side-channels during the device's design, to avoid vulnerabilities arising later on at the circuit or system level.

2. Circuit implementations that exhibit uniform power traces for all logic transitions, i.e., switching as well as non-switching, are particularly effective against PSC attacks, which rely on data-dependent power profiles to infer secret data. The presence of two different carrier types and related transistors (n and p) in CMOS essentially means that different inputs causing different logic transitions result in distinctive power profiles. However, some emerging technologies operate uniformly under various bias conditions and logic transitions, with respect to power consumption. For example, consider an emerging technology device with two different logic states, which are attained by applying a positive (negative) voltage bias to produce an equal amount of positive (negative) operational current, thus consuming the same power.

3. Particular emerging technologies are able to implement logic and memory more efficiently than CMOS. This results in smaller area and lower device count for the same function implemented. Besides, a common method for obfuscating the power profile is to leverage complementary functions while synthesizing the circuit, so as to make the power consumption balanced for every operation. However, the additional circuitry required for complementary functions renders this approach costly for CMOS logic. By operating at a fraction of the area and device cost, though, emerging technologies may afford for such a scheme.

8.3 Review of Selected Emerging Technologies and Prior Art

In this section, we briefly review prior applications of emerging technologies in mitigating side-channel attacks in electronic systems, particularly those focused on extracting information from the power characteristics of devices. The other side-channels of emerging technology-based logic and memory, like those targeting electromagnetic, photonic, or acoustic leakage, remain largely unexplored and are a topic of active research.

8.3.1 All-Spin Logic

The multi-functional all-spin logic (ASL) primitive described in Sect. 2.3.1 is leveraged by Alasad et al. in [AYL18] to demonstrate a PSC-resilient Advanced Encryption Standard (AES) application. As opposed to the input-dependent switching power in conventional technologies, ASL logic gates work on the principle of driving non-local spin signals to switch the output magnet, based on the spin polarization. The power dissipation in this process is uniform, irrespective of the initial and final states of the magnet. Particularly, the power consumption of ASL device remains approximately constant, independent of the supply voltage polarity, which is set according to the targeted final state of the magnet. This results in a uniform power profile. In other words, PSC-based attacks are hindered from inferring secret keys of an ASL-based cryptographic circuit since the spin current generated during the operation of a particular logic stage decides the final magnetization direction at its output stage. The magnitude of spin current required to orient the magnet in either direction remains the same. In contrast to ASL, other spintronic technologies like Magnetic Tunnel Junction (MTJ)-based logic and memories prove to be incompatible for implementing hardware crypto-systems owing to their differential power output and asymmetric read/write operations, which make them susceptible to PSC incursions [Kha+17].

To verify the hypothesized resilience against PSC attacks, Alasad et al. [AYL18] synthesize an AES circuit with ASL components. Note that ASL is also leveraged to implement the look-up tables (LUT) and shift registers required for the AES circuit, apart from the gates highlighted in Sect. 2.3.1. To alleviate the excessive power dissipation and delay issues of ASL circuits, a pipelined AES architecture is used, wherein buffers are inserted for path balancing and magnifying the spin currents, for improved and stable fan-out. A simple example circuit for pipelining is shown in Fig. 8.1. A CPA-based attack targeting the first round of the AES implementation, modeled using Hamming weight and Hamming distance, is shown to be successfully averted by substantially increasing the number of power traces required to disclose the secret key.

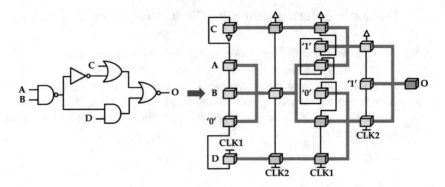

Fig. 8.1 Pipelined ASL architecture for an example circuit

8.3.2   Memristive Devices

Memristive devices operating in the deterministic regime, as opposed to the stochastic memristors studied in Sect. 5.3.3, are promising for non-volatile memory implementation (thanks to their ultra-low-power dissipation, high retention time, considerable endurance, low read and write latencies, and small area footprint) [YSS13]. The memristive element is particularly employed as the memory cell in resistive random access memory (RRAM) technologies.

For example, Khedkar et al. [KDK14] propose a DPA mitigation technique for crypto-processors, by obfuscating the power profile with complementary memristor-based RRAM modules. Their method attempts to equalize the power traces of AES such that every memory operation consumes the same amount of power, thus thwarting any inference from the power profiles. The corresponding architecture of the protected AES system is shown in Fig. 8.2. Here, the state memory is used to store the previous and next state of encryption/decryption. Khedkar et al. [KDK14] implement the state memory and the dual-state memory using RRAM, to minimize the overhead costs of having an additional complementary memory block. The power consumption in the memory banks is directly related to the data written into or read from the memory in each round. This dependency is exploited by DPA attacks, which observe the PSC leakage from the state memory connected to the AES. The presence of a dual-state RRAM, which stores (accesses) the complementary data with respect to the data stored (accessed) in the state RRAM, ensures that such dependency is avoided by balancing the power dissipation. For instance, if the original data being written into the state memory during any round is $(23)_{16}$, then the dual-state memory is written with $(DC)_{16}$, such that the overall power dissipation always corresponds to writing $(FF)_{16}$. The information about memory access and data written/read in the current round is obtained through a balancing logic module, which snoops the bus between the AES and memory unit. Although their technique is effective against conventional DPA attacks targeting the memory modules, it remains vulnerable to others like removal attacks which seek to disable the dual memory and/or the balancing logic altogether.

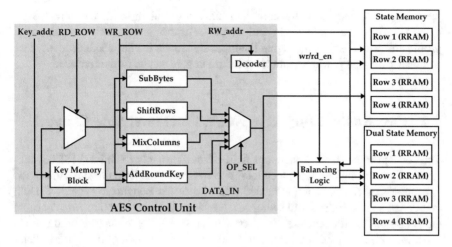

Fig. 8.2 DPA-resilient AES architecture with memristor-based dual RRAM

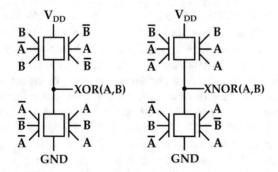

Fig. 8.3 Symmetric XOR/XNOR implementation with three-independent-gate SiNWFET, to obtain uniform power profiles for circumventing PSC attacks. The actual device structure of the SiNWFET is shown in Fig. 2.2

8.3.3   Silicon Nanowire Transistors

The complementary memory technique for smoothening power profiles, explored in [KDK14], can be extended to logic as well. However, as seen in Sect. 8.3.1, this poses a challenge in terms of area and power overheads, requiring up to twice the number of transistors as the original design. To overcome this, Giacomin et al. [GG18] propose the use of compact Silicon Nanowire FETs (SiNWFET) with three independent gates, to implement complementary logic that can decorrelate the power consumption of the circuit from the input signals.

Three-independent-gate SiNWFETs, by virtue of their two control gates and one polarity gate, are able to realize complex logic much more efficiently as compared to CMOS. For instance, XOR/XNOR gates based on these SiNWFETs require half the number of transistors as their CMOS counterparts (Fig. 8.3). This allows the designer to include extra complementary logic along with the actual logic to be protected, to equalize the output power traces and make the total power profile uniform. Also, by arranging the input signals on the three independent gates symmetrically, the power variation for the input gate can be further reduced.

8.4   Case Study I: Impact of NCFET on Power Side-Channel

The concept of negative capacitance field effect transistors (NCFETs) [SD08] represents a promising emerging technology to overcome the fundamental limit in the existing MOSFET technology, related to the so-called Boltzmann tyranny [SD08] that dictates the sub-threshold swing of a transistor to be above 60 mV/dec at room temperature. This fundamental limit restricts the scalability of the threshold voltage for new nodes, despite the innovations in transistor structures (i.e., from planar transistors to FinFET, then nanowire, even nanosheet structures). As a result, gains in performance and efficiency with technology scaling become increasingly harder to achieve for regular MOSFET technology. A point of inflection had occurred when GlobalFoundries demonstrated that NCFET could be made compatible with the CMOS fabrication process [Kri+17], by utilizing ferroelectricity in HfO_2-based materials, which are standard materials for the transistor dielectric.

Due to the presence of negative capacitance, the total gate capacitance of NCFET is always greater than the conventional/counterpart transistor [Amr+18]. Hence, at the same operating voltage, NCFET-based circuits consume higher switching power compared to traditional CMOS circuits. In turn, this necessitates a deeper understanding of security in the context of NCFET technology, in particular concerning PSCs. It is important to note that this represents a general challenge for emerging technologies—there is a need to explore devices during technology development also with particular focus on design aspects impacting hardware security. Ultimately, the emerging technologies should be quantified not only in terms of power, performance, and area, but also in terms of resilience against various attacks.

In order to assess hardware implementations of ciphers or other sensitive circuitry with respect to side-channel attacks, a widely adopted strategy is to implement them using a field-programmable gate array (FPGA). However, such an FPGA-based evaluation may not capture the resilience of an ASIC implementation accurately, due to the fundamental differences in the underlying circuit architectures, resulting in, e.g., differences in dynamic and static power consumption between FPGA-based and ASIC-based circuits; such differences have direct impact on PSC attacks. For most emerging technologies, like NCFET, FPGA-based evaluation is not even an option to begin with, due to the non-existence of such platforms like NCFET FPGAs before mainstream and high-volume manufacturing. Hence, CAD flows that allow developers to evaluate the resilience of their hardware for different technologies, already during design-time, become indispensable. Such a CAD flow supporting side-channel attack analysis will not only reveal the role that underlying technology plays, but it will also allow for: (1) the detection of vulnerabilities at design-time, providing an early feedback to improve the hardware implementation; (2) the accurate analysis of newly designed cryptographic hardware primitives and/or countermeasures.

In this case study, we describe the work by Knechtel et al. [Kne+20] in detail. The key idea is to enable device exploration for NCFETs and related parameter tuning early on, before manufacturing, yet with high-accuracy, based on well-characterized technology libraries and sophisticated CAD flow integration of the security assessment against power side-channel attacks. Accordingly, this case study addresses the two key challenges outlined above. The scope of this case study can also be summarized as follows:

1. The resilience of NCFET technology against PSC attacks is investigated for the first time. An attack-evaluation flow using commercial design tools for emerging technologies is also proposed and implemented.
2. The presented flow enables designers to analyze ciphers (or any other modules), running on emerging technologies (or established nodes), against the powerful CPA attack in particular. To showcase the flow, the resilience of the well-known AES is evaluated and compared for various NCFET technology setups characterized in [Kne+20].
3. Sophisticated NCFET standard-cell libraries and their counterpart baseline library (7 nm FinFET) are employed to analyze how the underlying technology impacts the power profile of the cipher hardware during operation. By doing so, the role played by the thickness of the ferroelectric (FE) layer in NCFETs with respect to PSC attacks, is studied. More specifically, the findings reveal that NCFET-based circuits are more resilient to the classical CPA attack, due to the considerable effect of negative capacitance on the switching power. The dependence of the higher resiliency of the NCFET-based circuits on the FE thickness is demonstrated, which opens new doors for optimization and trade-offs. However, note that this finding may not be generalized, as the success of CPA (or any attack, for that matter) depends on the accuracy of the leveraged device and power model. The presented flow can be easily used for other ciphers as well as for other emerging technologies, i.e., as long as standard-cell libraries for the technologies are available.

8.4.1 Power Consumption in NCFETs

For NCFETs, a thin layer of FE material is integrated within the transistor's gate stack. The FE layer behaves as a negative capacitance, resulting in a voltage amplification instead of a voltage drop as in conventional transistors. Such a higher internal voltage pushes the sub-threshold swing to move beyond its fundamental limit. This, in turn, has far-reaching consequences on the efficiency of circuits [Amr+18], but also on their resilience against PSC attacks; some details are explained next.

As indicated, the FE layer integrated within the transistor gate stack manifests itself as a negative capacitance that provides an internal voltage amplification inside the transistor ($V_{int} > V_g$). The voltage amplification (Eq. 8.1) is related to both the internal capacitance of the transistor ($C_{internal}$) and the negative capacitance obtained by the FE layer (C_{ferro}).

$$V_{int} = A_V \cdot V_g \; ; A_V = \frac{|C_{ferro}|}{|C_{ferro}| - C_{internal}} \tag{8.1}$$

To ensure no hysteresis: $|C_{ferro}| > C_{internal} \Rightarrow A_V > 1$

The internal voltage amplification magnifies the vertical electric field of the transistor, leading to a higher driving current. As can be seen in Fig. 8.4c, the drain current of a transistor in NCFET is always higher than the baseline current in the original CMOS FinFET. In fact, the thicker the FE layer, the higher the drain current. In their work, Knechtel et al. consider four different layer thicknesses of FE, namely 1 nm, 2 nm, 3 nm, and 4 nm, which they refer to as TFE1, TFE2, TFE3, and TFE4, respectively. The thickness is limited to 4 nm because for higher thicknesses, a hysteresis-free operation in NCFET transistor, which is essential to build logic gates, cannot be ensured any more [Amr+18]. The switching power ($P_{switching}$) of any logic is governed by the switching activity (α), total capacitance (C), operating voltage (V_{DD}), and operating frequency (f): $P_{switching} \propto \alpha C V_{DD}^2 f$. Because C_{ferro} exhibits a negative value and $|C_{fe}| > C_{int}$, the total capacitance of NCFET C_{NCFET} is, in fact, always greater than the FinFET baseline capacitance ($C_{internal}$):

$$C_{NCFET} = \frac{C_{ferro} \times C_{internal}}{C_{ferro} + C_{internal}} > C_{internal} \tag{8.2}$$

As Fig. 8.4d shows, NCFET transistors always exhibit a greater gate capacitance (C_{GG}). When the FE layer thickness is larger, the gate capacitance becomes higher due to the greater negative capacitance. Therefore, when a circuit is implemented in an NCFET technology, it will consume a higher switching power when compared to conventional CMOS technology. Thus, its susceptibility to PSC attacks might be different and requires detailed study.

8.4.2   CAD Flow for Power-Side Channel Evaluation

Next, the CAD flow by Knechtel et al. is discussed, which helps to evaluate cryptographic hardware against PSC attacks (Fig. 8.5). The authors demonstrate the flow using AES in their study; however, the flow is not limited to a specific cipher. The flow takes the register-transfer level (RTL) of the cipher circuit as input, along with the standard-cell library for the technology of interest, and it returns

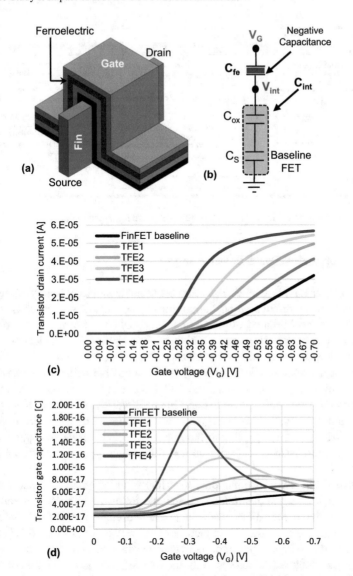

Fig. 8.4 (**a**) NCFET structure, with ferroelectric layer integrated inside the transistor's gate stack [Amr+18]. (**b**) Equivalent capacitance series, where the internal voltage exhibits a greater voltage ($V_{internal} > V_G$). (**c**) Drain current across gate voltage for a p-type transistor for four different NCFET cases compared to the FinFET baseline. (**d**) Large increase in the transistor's gate capacitance (C_{GG}) due to the negative capacitance effects

the minimum number of power traces required to break the cipher, i.e., to reveal the secret key. A greater number of power traces indicates a higher complexity for the attack and, hence, a lower susceptibility of the cipher hardware and/or the technology to PSC attacks.

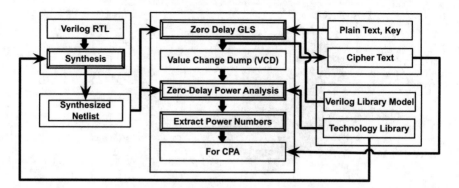

Fig. 8.5 Overview on CAD flow for PSC evaluation of emerging technologies

8.4.2.1 Simulation-Based Power Analysis

First, the cipher RTL is synthesized using the target technology library. Then, using a testbench, the functionality of the gate-level, post-synthesis netlist is verified. After the user specifies a set of plain-texts and a key (or set of keys), the testbench generates the corresponding set(s) of cipher-texts required for verification. During these gate-level simulations, a Value Change Dump (VCD) file is also generated; it captures the switching activity of every node within the netlist in a user-defined time resolution (e.g., 1 ps). Next, the VCD file is used for power simulation, along with the post-synthesis netlist and library/libraries of interest. Note that the dynamic power typically dominates the power consumption. This holds even more true for NCFET technology, due to the important role of negative capacitance for increasing the transistors' capacitance and thus magnifying the switching power; recall Sect. 8.4.1.

Instead of full timing simulations, Knechtel et al. leverage zero-delay simulations. The authors argue that doing so offers the following benefits. For zero-delay analysis, all signal transitions occur at the active edge of the clock, where the peak power values are easier to extract. For PSC evaluation, in fact, one is mainly interested in the switching power of particular registers; this switching power occurs always at the clock edge. Besides, only the relevant time intervals are considered, i.e., the last round of AES, which is known to be vulnerable when implemented in hardware [BCO04].

In short, the flow serves to obtain the design-time power traces for AES (or any other cipher) stepwise when processing the set(s) of texts for the secret key(s), and, for AES, peak power values are extracted for the related last-round operations. This evaluation has to be done separately for each technology setup, i.e., for NCFET with different FE configurations and for the FinFET baseline in this case study.

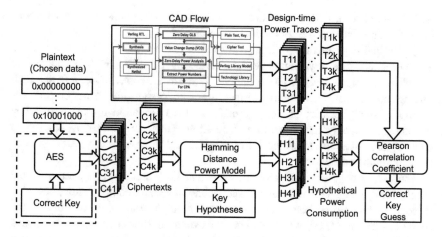

Fig. 8.6 Integration of CPA attack in the CAD framework

8.4.2.2 Correlation Power Analysis

The correlation power analysis (CPA) [BCO04] is a powerful attack that uses the Pearson correlation coefficient (PCC) to measure the relation between predicted and actual power profiles of a device undergoing some cryptographic operations. These operations are a function of variable data (i.e., the plain-/cipher-texts to encrypt/decrypt) and the secret key.

The integration of the CPA attack in the CAD framework is illustrated in Fig. 8.6. First, the predicted power consumption is derived from a power model [BCO04] as explained next. Note that such modeling is to be repeated for all possible candidates of the secret key; therefore, the resulting values are related to the key hypotheses and are referred to as hypothetical power values.

The dynamic power depends largely on switching activities, i.e., on rising signal transitions $0 \rightarrow 1$ and falling signal transitions $1 \rightarrow 0$. Typically, registers consume a significant portion of the dynamic power during such transitions. It is important to note that non-switching transitions (i.e., $0 \rightarrow 0$ or $1 \rightarrow 1$) do incur some power consumption as well, but on much smaller scales. Furthermore, there is also static or leakage power, which arises independent of switching activities. Thus, considering the Hamming distance (HD) for the output of registers before and after particular operations is well-established as HD power model [BCO04]. This model is simple yet effective, given that switching transitions impact both the dynamic power as well as HD of data directly, whereas non-switching transitions do not—these differences help for the actual correlation framework to infer which key hypotheses is correct, based on the VCD switching activities.

For the attack framework, the registers holding the intermediate round texts during the last-round operation are targeted to build up the HD power model. Importantly, this consideration is also in agreement with the scope of the collected design-time traces. Note that attacking the first or last round are both proven to be effective [BCO04], with the difference being the use of plain- or cipher-texts as known references required for the HD power model.

In the final step, the design-time power values and the hypothetical power values are correlated. The key candidates are sorted by the PCC, with the candidate exhibiting the highest PCC value being considered as the correct one. Note that such findings are subject to the number of measurements/design-time values considered; an overly small number of samples can bias the correlation toward wrong key hypotheses inducing the maximum correlation. Thus, an attacker is advised to conduct CPA over an increasing set of samples and to monitor the progression of the maximum correlation.

The above steps are sufficient for an actual CPA attack. For quantifying the susceptibility of the cipher hardware and technology under consideration, however, these steps have to be conducted throughout multiple trials, e.g., by varying the secret key and the texts, while tracking the progression of success rate (i.e., runs where the secret key is correctly inferred over all runs); see Sect. 8.4.3.1 for further details.

8.4.3 Experimental Evaluation

8.4.3.1 Setup for CAD Flow and Correlation Power Analysis

For the implementation of the CAD flow, Knechtel et al. employ commercial tools as follows. *Synopsys VCS M-2017.03-SP1* is used for functional simulations at RTL and gate level, *Synopsys DC M-2016.12-SP2* for logic synthesis, and *Synopsys PrimeTime PX M-2017.06* for power simulations.

For the AES implementation, a regular RTL is leveraged that is working on 128-bit keys and 128-bit texts, using look-up tables for the AES substitution box, and without using any PSC countermeasures. The AES netlist is provided in [Kne20]. For the AES circuit implementations using the FinFET baseline technology as well as all NCFET-specific technology setups, the same supply voltage $V_{DD} = 0.7V$, the same switching activities α (as dictated by *VCS*), and the same frequency $f = 100$ MHz are used. Thus, $P_{switching}$ differs only for varying capacitances C, given that $P_{switching} \propto \alpha C V_{DD}^2 f$.

For the CPA implementation, Knechtel et al. use and extend an open-source C/C++ framework; the authors provide their version in [Kne20]. All CPA runs are executed on a high-performance computing (HPC) facility, with 14-core Intel Broadwell processors (Xeon E5-2680) running at 2.4 GHz, and 4 GB RAM are guaranteed (by the Slurm HPC scheduler) for each CPA run. Note that, to manage the complexity of the CPA attack, the correlation between design-time power values and the hypothetical power values is evaluated stepwise on the bytes of the last-round text, not the whole text at once. Thus, instead of having to consider all $2^{128} \approx 3.4 \times 10^{38}$ possible key candidates, which is computationally intractable, only the resulting $16 \times 2^8 = 4096$ key candidates are considered. Therefore, the key hypotheses are also built up byte by byte, considering the respective key-candidate bytes exhibiting the highest PCC values. While doing so limits the accuracy for the correlation-based analysis in theory [Fei+15], it is common practice and still sufficiently accurate in practice.

For the CPA trials, Knechtel et al. generate and store ten random keys (128 bits each). The authors also generate and store 2000 random plain-texts (also 128 bits) and, separately for each key, the corresponding 2000 cipher-texts. The authors keep the sets of keys and corresponding texts the same across the technologies setups; hence, the simulated power values are guaranteed to vary only due to the underlying technology. Finally, to enable the CPA attack to progress stepwise through the sample space of all power traces (to thoroughly quantify the success rate depending on the number of traces considered), Knechtel et al. generate and store batches of permutations for power values and corresponding texts as follows. Over the course of 1000 steps, the authors randomly pick 1000 sets of power values/texts for each step such that each step describes 1000 permutations in multiples-of-two of all 2000 available pairs of power values/texts. In total, this results in 1,000,000 permutation sets of increasing size.[1] Knechtel et al. independently generate three such batches of 1,000,000 permutation sets, which are employed for three independent CPA trials. To ensure fair comparisons, it is essential that these batches are all memorized, stored, and re-applied when conducting the CPA attack for different keys as well as for different technology setups.

8.4.3.2   Setup for NCFET-Aware Cell Library Characterization

As explained above, a standard-cell library of the target technology is essential. In their work, Knechtel et al. employ the open-source 7 nm FinFET cell library as baseline technology [Cla+16]. Then, the authors characterize the library to create the required NCFET-specific cell libraries [Amr+18] (Fig. 8.7). To achieve that, the authors use SPICE simulations for post-layout netlists including parasitic information for a wide range of sequential and combinational standard cells provided within the 7 nm PDK [Cla+16].

As is typically done in any commercial standard-cell library, Knechtel et al. characterize power and delay of every standard cell under 7 input signal slews and 7 output load capacitances. To model the impact of FE layer and the negative capacitance effect on the electrical characteristics of nMOS and pMOS transistors, the authors employ the state-of-the-art physics-based NCFET model [Pah+16]. They integrate the model in a self-consistent manner within BSIM-CMG, which is the industry standard compact model of FinFET technology [Amr+18]. The material properties, required for the NCFET physics-based compact model, were obtained from the experimental measurement presented in [Mül+12].

---

[1] For example, for the first three steps, the authors might randomly select the following sets of power values/texts permutations, encoded by indices in the range 1–2000: {734, 1297}, {87, 815, 562, 33}, and {245, 734, 12, 1395, 1553, 94}.

Fig. 8.7 NCFET-specific library characterization for 7 nm FinFET technology

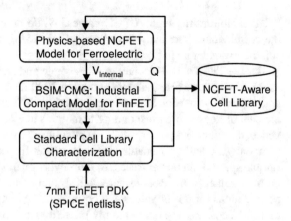

For a comprehensive analysis, recall that Knechtel et al. consider four different cases for FE thicknesses, ranging from 1 nm to 4 nm and referred to as TFE1, TFE2, TFE3, and TFE4, respectively. Their cell libraries are fully compatible with commercial CAD tools. Therefore, the authors can directly deploy them within their flow to investigate the resiliency of AES (or other ciphers) against PSC attacks when NCFET technology is used in comparison to the FinFET baseline technology.

8.4.3.3   Analysis and Insights

The primary goal of the study by Knechtel et al. is to investigate the resilience of the NCFET technology against PSC attacks, i.e., the classical CPA attack in particular. The key finding of their study is the following: the thicker the FE layer, the more resilient the device becomes. This is because for thicker FE layers, the negative capacitance effects become greater and, thus, the dynamic power becomes more dominant but also more varied for the different switching transitions, rendering the classical CPA more difficult. It is essential to note that it may not be possible to generalize these findings. This is because the observed resilience is w.r.t. the classical CPA using a HD model. While this particular attack setup is a powerful and state-of-the-art approach, other attacks may lead to other findings as well. This further implies that, as is most often the case for security, a thorough evaluation of the various possible attack models is required before deriving final conclusions for guiding the implementation of sensitive circuitry. Next, we review the findings by Knechtel et al. in more detail.

In Table 8.1, the number of traces required for the CPA attack to infer the correct key for varying success rates are disclosed. For example, for a 50% success rate, 500 out of 1000 runs provide the correct key. The authors report the averages and the variations (i.e., standard deviations) across all ten random-but-reproducible keys, and they do so for three independent trials employing the scheme of random-but-reproducible permutations for power values/cipher-texts. Figure 8.8 illustrates the

Table 8.1 Statistics on traces required on average for successful CPA runs

Technology setup	Trial 1		Trial 2		Trial 3	
	Avg.	Std. Dev.	Avg.	Std. Dev.	Avg.	Std. Dev.
50% Success rate						
FinFET	501.5	46.8	502.0	46.6	502.2	45.7
TFE1	503.4	47.2	505.6	48.3	504.3	47.3
TFE2	509.5	49.7	509.6	52.5	507.8	46.2
TFE3	519.6	52.2	520.4	48.3	519.6	51.4
TFE4	547.8	58.9	546.0	57.3	545.6	56.2
90% Success rate						
FinFET	693.0	90.3	688.8	80.6	693.6	86.4
TFE1	698.4	93.5	696.2	92.6	695.2	88.1
TFE2	706.8	94.8	702.2	94.7	699.6	87.9
TFE3	724.4	107.3	718.8	101.2	721.6	103.8
TFE4	759.2	117.4	758.6	111.8	760.8	122.1
99.9% Success rate						
FinFET	985.2	156.9	991.4	148.9	983.0	151.8
TFE1	987.6	156.2	988.0	144.3	982.6	151.9
TFE2	995.2	151.5	999.2	159.9	985.0	158.7
TFE3	1011.6	156.4	1026.6	164.7	1013.4	165.9
TFE4	1066.0	192.3	1065.8	185.9	1068.4	187.8

For a fair comparison, the same selection of traces is considered across all technology setups. Results are obtained considering all ten random-but-reproducible keys and for three independent trials employing the scheme of random-but-reproducible permutation of traces

progression of success rates over CPA steps (i.e., traces considered), again averaged over the same ten keys, and further averaged over the same three trials. This way, Knechtel et al. enable a truly fair and robust comparison across all technology setups.

Knechtel et al. conclude from both Table 8.1 and Fig. 8.8 that TFE4 provides the most resilient setup and the FinFET baseline provides the weakest setup, respectively. More specifically, TFE4 is on average 8.13% more resilient than the FinFET baseline (i.e., for a success rate of 99.9%). The authors further conclude that the increase of both the average resilience and the variation of resilience from FinFET baseline to TFE4 is not linear, but more pronounced toward TFE4. This is because, as demonstrated in Fig. 8.4d, a larger thickness of the FE layer results in a higher, non-linear increase in the transistor gate capacitance; the most significant increase occurs for TFE4. As explained earlier, the total dynamic power (leveraged by the CPA framework) depends on the capacitances across the whole circuit. In fact, the total capacitances of standard cells and the related impact on dynamic power become both more dominant and more varied for thicker FE layers, as explained next.

Despite the FE layer being identical for the nMOS and pMOS transistors in all the standard cells, the related negative capacitance effects play out differently. This

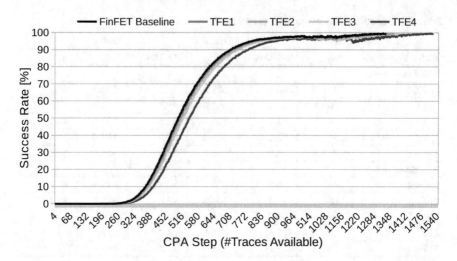

Fig. 8.8 CPA progression in terms of success rates over traces being available for consideration. Success rates are averaged across all ten random-but-reproducible keys and across three independent trials considering the random-but-reproducible permutation sets for power values and cipher-texts; thus, there are 30,000 CPA runs underlying for each data point for all curves. This ensures a fair and robust comparison across all technology setups, along with a thorough sampling across an increasing number of traces being available for consideration in the CPA framework

is essentially because of different capacitance matching between the FE layer added to the transistor gate and the MOS capacitance of the underlying transistor and, hence, varying differential gains in nMOS and pMOS transistors. The differential gain (A_V) and the resulting average gain (A_{avg}) are given in Eq. 8.3 and related simulation results shown in Fig. 8.9. As can be observed, nMOS transistors exhibit a higher average gain than pMOS transistors.

$$A_V = \frac{\partial V_{int}}{\partial V_G}, \text{ where } A_{avg} = \frac{1}{V_G} \int_0^{V_G} A_V \, dV_G \qquad (8.3)$$

Such dis-proportionality leads, in turn, to a varying impact on the switching power for nMOS and pMOS transistors. Accordingly, Knechtel et al. observe that the switching power for rise versus fall transitions of the AES round-operation registers varies to a greater extent for TFE4 when compared with the FinFET baseline (Table 8.2). The varying impact which the negative capacitance plays for pMOS versus nMOS transistors, especially for thicker FE layers, thus represents the root-cause for the higher CPA resilience observed for the NCFET technology—the greater the variations for the rise versus fall switching power are, the more "noisy" the power samples become, and the lower the accuracy becomes for the classical CPA based on the well-established HD power model.

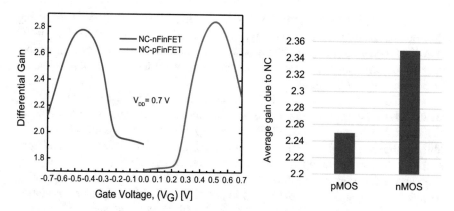

Fig. 8.9 Differential gain due to negative capacitance in nMOS and pMOS FinFET transistors (left) and the corresponding average gain (right)

Table 8.2 Power values for AES round-operations registers

Power components		FinFET baseline	TFE4
Static or leakage		$3.45E - 10$	$3.17E - 10$
$0 \rightarrow 1$ transitions	Clk rise	$4.89E - 07$	$6.94E - 07$
	Switching	$1.70E - 06$	$3.89E - 06$
	Total	$2.19E - 06$	$4.59E - 06$
$1 \rightarrow 0$ transitions	Clk rise	$4.89E - 07$	$6.94E - 07$
	Switching	$1.55E - 06$	$3.17E - 06$
	Total	$2.04E - 06$	$3.86E - 06$
$1 \rightarrow 1, 0 \rightarrow 0$	Clk rise	$4.89E - 07$	$6.94E - 07$
transitions	Total	$4.89E - 07$	$6.94E - 07$

8.4.3.4   Discussion

In Fig. 8.8, note that success rates are dropping momentarily toward the end of the curves. This is because the longer the CPA attack progresses, the fewer keys remain unresolved, and those remaining "resilient keys" tend to incur lower success rates over longer periods, resulting in these momentary drops. Such varying resilience across keys is also observable from the standard deviations reported in Table 8.1. Knechtel et al. caution that this finding should not motivate one to favor specific, seemingly "resilient keys" for actual applications—such outcomes are highly dependent on the architecture, RTL, and gate-level implementation of the cipher, as well as the selection of plain-/cipher-texts. Even for very same sets of texts, which Knechtel et al. have used for all experiments (to enable a fair comparison), the authors argue that they naturally expect and have indeed observed varying distributions for the bit-level flips across the last-round texts of the AES cipher when using different keys. These varying distributions, among other aspects, can impact the accuracy of the CPA, given the byte-level working of the classical HD power model [Fei+15]. Knechtel et al. also note that other studies are investigating the role of keys and texts more thoroughly, e.g., based on collision, confusion, and/or statistical modeling [Fei+15]. In any case, their findings on NCFET hold given that they are centered on average results obtained across all ten keys as well as across three independent CPA trials.

As indicated before, the flow by Knechtel et al. is applicable to any technology, be it CMOS, NCFET, or other emerging technologies, as long as those devices are supported by commercial CAD flows, especially by means of standard-cell libraries being available. While one important aspect for different technologies is their different operating frequencies, also recall that CPA considers only the peak power for relevant registers' transitions. Therefore, while different operating frequencies will scale the power amplitude/values, it will not hinder correlation within the CPA framework. In fact, Knechtel et al. have conducted a selection of the same experiments underlying Table 8.1 for various frequencies, and they have observed the same number of traces for all configurations.

While the absolute numbers of traces required and their differences across setups may seem small to some readers (in particular those familiar with PSC attacks on CMOS devices in the field), the related findings are significant nevertheless. As they are grounded on physics-based compact models and thoroughly characterized standard-cell libraries, along with a zero-delay power simulation using commercial-grade tools, these "ideal case" findings represent firm boundaries for any future CPA attack on NCFET devices in the field. That is, given the inevitably noisy behavior of future devices in the field, an attacker cannot perform any better than what Knechtel et al. observed—assuming the same operating conditions. Thus, from a security-enforcing designer's perspective, these findings can serve well as conservative guidelines, e.g., for security schemes like dynamic key updating [TS15].

As indicated, the root-cause for the observed resilience of the NCFET technology against the CPA attack are the variations for rise versus fall switching power being more pronounced for thicker FE layers. The root-cause and the resulting effect of power samples becoming more "noisy" can be expected to have a detrimental impact on other PSC attacks as well, but related quantitative studies are still outstanding.

8.5 Case Study II: Impact of 3D Integration on Thermal Side-Channel

3D integration is expected to successfully meet the increasingly demanding requirements for modern ICs, such as high performance, increased functionality, and low power consumption. In fact, various studies, prototypes, and commercial products have shown that such 3D and 2.5D ICs offer significant benefits over conventional 2D ICs [Fic+13, Iye15, Kim+12, Shi18, PSF17, P D+17, Kim+19, Sto+17, Cle+16, Tak+13, Lau11, Viv+20]. See also Sect. 6.1 for more background on the emerging 3D and 2.5D IC technologies.

Various SCAs on ICs have been successfully demonstrated and also mitigated at least to some degree (see also Sect. 1.1.1.2). In the context of emerging 3D ICs, however, SCAs have only recently come into focus. Conceptual studies such as [Xie+16] discuss how structural properties of 3D ICs may help in mitigating SCAs. It is intuitive that SCAs developed for 2D ICs may not be successful in 3D ICs, due to the different physical structures of 2D and 3D ICs. For example, the authors of [DY18] studied power side-channel attacks on 3D ICs, where they observed that the noise profiles from the stacked chips within the 3D IC are superposed, rendering related attacks more difficult in general.

One major concern for 3D integration in general and for 3D logic ICs in particular is thermal and power management, which is furthermore highly technology-dependent. The latter is because the following all play important roles [KL16, Pen+15, Sam+16, Cou+16]: (i) the stacking of multiple substrates and their active layers, along with thermal interface materials between, (ii) the 3D interconnects like through-silicon vias (TSVs) or face-to-face (F2F) microbumps along with large ranges of feasible pitches and arrangement patterns, and considered jointly as (iii) the variation of heat dissipation and distribution of power within the active layers and their 3D interconnect networks. In practice, 3D ICs will likely require some runtime capabilities for thermal management, e.g., based on embedded on-chip thermal sensors [Ben+16, Zhu+08, Fu+17]. Not only but also considering the availability of such sensors, a practical and easy-to-exploit SCA could be based on thermal readings of 3D ICs, resulting in the thermal side-channel.

The thermal side-channel (TSC) in general has gained attention thanks to case studies such as [HS14, Mas+15, INK11]. For example, Masti et al. [Mas+15] have shown for Intel Xeon multi-core processors that (i) process executions on one core can be detected in adjacent cores, and that (ii) different processes, when scheduled by turns in one core, can build a covert channel with up to 12.5 bit/s. Still, there are also practical limitations of the TSC, like limited bandwidth or assumption of known and linear-behaving heat paths; see Sect. 8.5.1.1 for more details. Given that 3D ICs represent highly integrated, stacked, and complex interconnects electronics by nature, it remains an interesting question whether the TSC may be particularly leaky or not in 3D ICs.

In this case study, we describe the related work by Knechtel and Sinanoglu [KS17] in detail. The key idea of the authors is to carefully exploit the specifics of material and structural properties of 3D ICs, seeking to decorrelate the thermal patterns from the underlying power and activity patterns, in order to mitigate the TSC leakage (Fig. 8.10). The scope of this case study is summarized as follows:

1. Security is considered early on, as a design parameter. More specifically, an SCA-aware floorplanning methodology that continuously tracks the correlation and entropy of thermal patterns within the 3D IC, is demonstrated.
2. Floorplanning is considered as target for the security-centric design of 3D ICs for the following two reasons. First, due to the high integration density of 3D ICs, optimizing the thermal and power distributions is essential, with floorplanning widely acknowledged as appropriate design stage to do so [Lim13, KL16,

Fig. 8.10 The thermal
side-channel can be mitigated
in 3D ICs by their design,
specifically the arrangement
of TSVs helps to induce
disturbances for the
heat-conduction paths

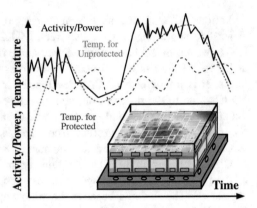

Van+11, XC17]. Second, designers typically have to reuse at least some hard
intellectual-property (IP) modules, with limited access to only basic properties
like area, aspect ratio, inputs/outputs, etc., while they still hope to mitigate
potential SCAs. The methodology proposed by Knechtel and Sinanoglu is
applicable in such scenarios through leakage-aware design techniques during
block-level floorplanning.

3. An attacker can monitor the thermal patterns of an IC either at regular runtime
   or when he/she applies specific input patterns, crafted to reveal the activity of
   sensitive components. Both scenarios for the attacker's capabilities and the SCAs
   they formulate and evaluate, are addressed.
4. The work is made publicly available within the open-source 3D floorplanner
   Corblivar [Kne16].

8.5.1   Thermal Side-Channel

8.5.1.1   Fundamentals

Among the different side-channels being exploited within various attack studies, the
TSC can be particularly attractive for three reasons: (i) it is easy to access via widely
available on-chip sensors [Mas+15, Ben+16, Zhu+08, Fu+17] or via thermal
readings from the outside, albeit the latter exhibits limited accuracy; (ii) it provides
internal and external leakage of activity/computation patterns through thermal
variations [HS14, Mas+15]; (iii) it may serve well as proxy for the power side-
channel using temperature-to-power interpolation techniques such as [Pae+13]. An
attacker may obtain localized thermal readings either directly from a network of on-
chip sensors (if available) or estimate them, e.g., using approximative interpolation

techniques leveraging few sensors. Note that interpolation techniques have been demonstrated to achieve high-accuracy and high-resolution estimates of both power and temperatures in large-scale 3D ICs [Ben+16].

Still, there are also practical limitations of the TSC to be considered, namely the following:

- The classical heat equation models power and temperature as linearly correlated. In practice, however, this behavior is constrained by material properties, i.e., thermal conductivities and capacitances both within and across the boundaries of chips [HS14]. In other words, a linear correlation of (secret) activity patterns and (leaked) thermal patterns is only applicable for homogeneous material properties. This has direct implications for the proposed study, as elaborated in Sect. 8.5.1.2.
- Due to relatively large thermal capacitances and resistances present in modern ICs, especially in 3D ICs, the internal heat flow is much slower than the underlying activity/power consumption patterns (see also Fig. 8.10). That is, the TSC has a relatively low bandwidth, which hinders the leakage of highly dynamic computation [HS14]. Knechtel and Sinanoglu assume strong capabilities for practical attacks (Sect. 8.5.2.2), rendering the TSC attractive nevertheless.
- As with any side-channel, the TSC experiences noisy readings; the heating effects of active modules are both spatially and temporally superposed. For any 2D/3D IC, these noise patterns may range from negligible to dominant effects, depending on the floorplan, the resulting power distribution, the material properties, and the heat paths along with their dissipation capabilities. Knechtel and Sinanoglu's assumptions for the attacker's capabilities address and mitigate also this limitation (Sect. 8.5.2.2).

8.5.1.2 Thermal Side-Channel in 3D ICs

Prior Art The work of Gu et al. [Gu+16] seeks to mitigate thermal-related leakage in 3D ICs, and is of particular interest here. The authors propose to integrate thermal sensors along with the actual functional circuitry, thermal-noise generators, and runtime controllers. These primitives inject dummy activities when/wherever considered necessary for smoothing of thermal profiles, i.e., to hinder thermal profiling of module activities. However, there are major shortcomings of this work. First, the concept appears not specifically tailored for 3D ICs—it does not explicitly address the different structural and material properties, different heat paths, or thermal coupling effects in 3D ICs, nor does it elaborate on the implementation of 3D ICs. Second, the principle of injecting additional activities causes further power dissipation, which can be prohibitive for thermal- and power-constrained 3D ICs in the first place. This shortcoming is confirmed by an observation in [Gu+16] itself: the best leakage-mitigation rates are only achievable for the highest injection rates.

Exploration First, we discuss the exploratory experiments conducted by Knechtel and Sinanoglu for understanding the TSC in TSV-based 3D ICs using two dies. It is important to note that thermal maps would be considerably different for other

3D integration styles, e.g., for monolithic 3D ICs [Sam+16]. Still, the proposed leakage-mitigation techniques are generic; they may be tailored toward different styles.

For the thermal analysis, Knechtel and Sinanoglu use *HotSpot 6.0* [ZSS15]. They model the heatsink atop the 3D IC, the signal TSVs acting as "heat-pipes" between stacked dies, and also the secondary path conducting heat toward the package. Further details are given in Sect. 8.5.3. Knechtel and Sinanoglu investigated all 30 combinations of 5 different power distributions (globally uniform, locally uniform, medium gradients, small gradients, and large gradients) and 6 different TSV distributions (no TSVs; maximal TSV density, i.e., 100% of area covered by TSVs and their keep-out zones; irregular TSVs; irregular TSVs along with regular TSVs; irregular groups of densely packed TSVs, i.e., TSV islands; and TSV islands along with regular TSVs). Note that some of these power and TSV distributions are impractical, yet relevant for exploratory experiments. Figure 8.11 illustrates a few selected power and thermal maps.

The key initial finding for thermal-leakage trends in 3D ICs is that the correlation of activity/power patterns and the thermal behavior mainly depends on both (i) the power-density distributions and (ii) the TSV distributions. More specifically, Knechtel and Sinanoglu observe high correlations for: (i) non-uniform power distributions with large gradients, both within dies and across dies and (ii) large numbers of regularly arranged TSVs. Further details are explained next.

1. *High TSC leakage for non-uniform power distributions with large gradients, both within dies and across dies.* For example, Fig. 8.11e–h exhibits large power gradients, while Fig. 8.11a–d and i–l show somewhat smooth gradients. One can observe the lowest correlations for globally uniform power distributions (Fig. 8.11a–d), which are, however, artificial and impractical. Locally uniform power distributions are more realistic, and one can observe that such distributions exhibit low correlations as well, especially along with irregular TSVs or distributed TSV islands. For example, comparing Fig. 8.11k,l with Fig. 8.11c,d, one can observe largely decorrelated thermal patterns, although the underlying power distribution of the former appears notably more diverse than its counterpart.

2. *High TSC leakage for large numbers of regularly arranged TSVs.* For example, see Fig. 8.11e–h with regularly placed TSVs and TSV islands versus Fig. 8.11a–d with irregularly placed TSVs and Fig. 8.11i–l with only TSV islands. Note that the use of few regular TSVs and/or TSV islands may locally further reduce the correlation. For example, see Fig. 8.11j, where a local thermal minima is observable in the lower right region, despite the relatively high power density underlying (Fig. 8.11i). It may be more intuitive to formulate these findings vice versa—the less regular and/or fewer the TSVs, the lower is the correlation. As indicated for the limitations of the TSC (Sect. 8.5.1.1), an unbiased and high correlation is only found in homogeneous structures. Inserting copper/tungsten-based TSVs into the silicon dies invalidates that assumption, especially when TSVs are few and irregularly distributed.

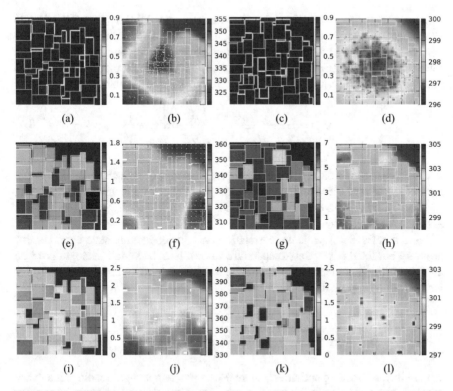

Fig. 8.11 A 3D IC floorplanned on two dies, with three different power scenarios illustrated in rows. From left to right, each row illustrates the power-density map of the lower die (**a, e, i**), the thermal map for that die (**b, f, j**), and the power (**c, g, k**) and thermal map (**d, h, l**) for the upper die. Each scenario exhibits distinct patterns for power and TSV distributions (the latter are illustrated as white dots in [**b, f, j**]), resulting in different trends for the power-temperature correlations as follows. Top row: artificially unified power for all modules with irregularly placed TSVs, resulting in the lowest correlation; middle row: large power gradients with both regular TSVs and TSV islands, resulting in the highest correlation; bottom row: groups of locally similar power regimes with TSV islands, resulting in relatively low correlation. The units of the power and thermal maps are $10^{-2} \mu W / \mu m^2$ and K, respectively

8.5.2 Methodology

8.5.2.1 Models for Thermal Leakage

Correlation and Correlation Stability Knechtel and Sinanoglu propose the Pearson correlation of power and thermal maps as a key metric for thermal-related leakage. Note that the Pearson correlation is also the underlying measure for the side-channel vulnerability factor (SVF) [Dem+12], an established metric for the vulnerability of ICs in terms of information leakage via side-channels.

Along with the assumptions for the attacker's capabilities (Sect. 8.5.2.2), the Pearson correlation is comparably meaningful as the SVF. That is, the lower the correlation, the lower the leakage of power/activity patterns via the TSC, and the lower the vulnerability of the IC. Given the power and thermal maps of a 3D IC, the authors measure the correlation coefficient r_d on each die d separately:

$$r_d = \frac{\sum_{i=1}^{n}(p_i - \bar{p})(t_i - \bar{t})}{\sqrt{\sum_{i=1}^{n}(p_i - \bar{p})^2}\sqrt{\sum_{i=1}^{n}(t_i - \bar{t})^2}} \tag{8.4}$$

where $\bar{p}, \bar{t}$ are the average power and temperature values, respectively, and p_i, t_i are the individual power and temperature values, both over all n locations in die d. The locations/values are to be organized in grids with same dimensions for both power and thermal maps/grids.

Note that Eq. 8.4 models the correlation only for one steady-state case, i.e., for arbitrary but fixed power and temperature values. Knechtel and Sinanoglu propose another correlation measure to capture the runtime stability of correlation $r_{d,x,y}$ for locations/bins x, y on die d:

$$r_{d,x,y} = \frac{\sum_{i=1}^{m}(p_{i,x,y} - \overline{p_{x,y}})(t_{i,x,y} - \overline{t_{x,y}})}{\sqrt{\sum_{i=1}^{m}(p_{i,x,y} - \overline{p_{x,y}})^2}\sqrt{\sum_{i=1}^{m}(t_{i,x,y} - \overline{t_{x,y}})^2}} \tag{8.5}$$

with m different sets of activities, where power and temperature readings may vary notably. The lower the stability $r_{d,x,y}$, the lower the thermal leakage for various ranges of power/activity patterns at particular locations x, y within the die d of the 3D IC.

Spatial Entropy of Power Maps As proposed by Claramunt [Cla05], the spatial entropy assesses the dispersion of the classical entropy over some regions. The definition of the spatial entropy is based on two principles: (i) the closer the different entities, the higher the spatial entropy; and (ii) the closer the similar entities, the lower the spatial entropy. Knechtel and Sinanoglu note that this is in good correspondence to the phenomenon of heat distribution—the closer the differently powered heat sources, the higher the spatial entropy of the superposed thermal responses, i.e., the thermal gradients; and the closer the similarly powered heat sources, the lower the thermal gradients.

Hence, the notion for modeling the spatial entropy of power maps is to anticipate the degree of the thermal gradients; recall that large thermal gradients hint on large thermal leakage (Sect. 8.5.1.2). The authors' formulation, derived from [Cla05], measures the spatial entropy of power maps in 3D ICs for each die d as follows:

$$S_d = -\sum_{i=1}^{n} \frac{d_i^{\text{inter}}}{d_i^{\text{intra}}} \left(\frac{|c_i|}{|C|} \log_2 \frac{|c_i|}{|C|}\right) \tag{8.6}$$

where $c_i \in C$ are classes of similar-value power ranges, and d_i^{inter} and d_i^{intra} are the average spatial inter- and intra-class distances. For all classes $c_i \in C$, where $|c_i| > 1$ and $c_i \neq C$, the authors calculate those distances d_i^{inter} and d_i^{intra} over all its values according to [Cla05]:

$$d_i^{\text{inter}} = \frac{1}{|c_i| \times (|c_i| - 1)} \sum_{j \in c_i} \sum_{\substack{k \in c_i \\ k \neq j}} dist(j, k) \tag{8.7}$$

$$d_i^{\text{intra}} = \frac{1}{|c_i| \times |C - c_i|} \sum_{j \in c_i} \sum_{k \in (C - c_i)} dist(j, k) \tag{8.8}$$

where $dist(j, k)$ are Manhattan distances for bins (i.e., class members) in the equidistant power-map grids constructed for each die.

For fast and effective classification of those grids, Knechtel and Sinanoglu employ the principle of nested-means partitioning: The power values are first sorted, then recursively bi-partitioned with the current mean defining the cut, and the partitioning proceeds until either a minimal size for any partition/class is reached or the standard deviation within any class approaches zero.

As the spatial entropy does not account for thermal analysis and subsequent correlation calculations, it lacks the capability to verify the actual leakage. It is, however, suitable for fast estimation of the potential leakage during floorplanning loops/iterations. During their experiments (Sects. 8.5.1.2 and 8.5.3), Knechtel and Sinanoglu observe the following trend for the bottom die ($d = 1$), even for different TSV patterns: the lower the spatial entropy, the lower the power-temperature correlation.

8.5.2.2 Threat Model

Assumptions An attacker has direct and physical access to the targeted 3D IC, but can conduct only non-invasive attacks.[2] That is, he/she may apply (arbitrary) input patterns and observe both the actual outputs and the thermal behavior of the 3D IC. It is also assumed that the attacker has a system-level understanding of the 3D IC, e.g., as obtained from datasheets. This is important for purposefully crafting input patterns to trigger certain activities.

Knechtel and Sinanoglu make further strong assumptions enabling an attacker to circumvent the critical limitations of the TSC (Sect. 8.5.1.1). First, the attacker can stabilize the 3D IC's activity with the help of specifically crafted, repetitive input patterns. Second, he/she may await the thermal steady-state response after

---

[2] In contrast to 2D ICs, one may expect invasive probing and reverse engineering of 3D ICs to be notably more difficult [KPS19a].

applying any input. Both assumptions help an attacker to ensure that the TSC readout correlates as best as possible with the input patterns. Third, the attacker has unlimited access to all thermal sensors, spread across the multiple dies of the 3D IC, and can thus obtain high-accuracy and continuous thermal readings of any (part of a) module at will.

Knechtel and Sinanoglu argue that these assumptions are relevant and practical when attacking security-critical (3D) ICs. For example, a security module may check whether a provided password is correct, and only then trigger data decryption. The thermal patterns for complex decryption operations will be relatively easy to distinguish from simple matching operations for password checks. As a result, an attacker may then brute-force a password even when the security module itself is "silent", i.e., when it would not directly react to wrong passwords.

Attack Scenarios Attacks exploiting the TSC in 3D ICs are formulated next, with consideration of the attacker's capabilities outlined above. Note that similar attacks have been discussed and successfully conducted on classical (2D) microcontrollers by Hutter and Schmidt [HS14].

1. *Thermal characterization of the 3D IC:* This is typically an exploratory attack, and others may follow as outlined below. Step by step, the attacker will apply a broad and varied range of input patterns in order to trigger as many activity patterns as possible. By monitoring the TSC, he/she can then build a model for the thermal behavior of the 3D IC.
2. *Localization and monitoring of modules:* The attacker targets on particular modules by applying crafted input patterns; the objective is to trigger these modules and observe thermal variations exclusively or at least predominantly within these modules. Then, the TSC is monitored for notable thermal patterns resulting from applying those input patterns. Given the attacker's lack of implementation details and hence the lack of a precise input-to-activity mapping, this attack is typically applied iteratively, with varying inputs. In case the characterization attack outlined above has been successfully applied; those efforts may be much lower. Once the thermal response is confined to particular regions, i.e., modules of interest are localized with some confidence, further attacks may be applied. Most notably, an attacker may now observe the sensitive activity/computation of particular modules by monitoring them during runtime.

8.5.2.3   CAD Flow

The objective of the work proposed by Knechtel and Sinanoglu is to account for thermal leakage in 3D ICs *directly and continuously within the floorplanning stage* that also optimizes for other design criteria. That is, during their iterative floorplanning flow (Fig. 8.12), the authors memorize the 3D floorplans with lowest correlations coefficients and spatial entropies (Sect. 8.5.2.1) as best in terms of thermal leakage, but select the final solution while also accounting for other criteria such as wirelength and critical delays. See Sect. 8.5.3 for setup details.

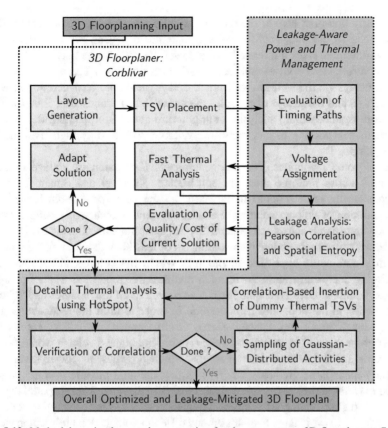

Fig. 8.12 Methodology, implemented as extension for the open-source 3D floorplanner *Corblivar* [Kne16, KYL15]

Knechtel and Sinanoglu implement and publicly release all their techniques within *Corblivar* [KYL15, Kne16]. They opt for this open-source 3D floorplanning framework mainly because it is multi-objective, modular, and competitive. An essential feature of *Corblivar* is its fast thermal analysis, which enables to continuously estimate the correlation for this work. However, the authors found this fast analysis to be inferior to the detailed analysis of *HotSpot* [ZSS15], especially for diverse arrangements of TSVs. Thus, the authors also verify the final correlation after floorplanning.

Management of Power Distributions Recall that non-uniform power distributions with large gradients induce notable thermal leakage (Sect. 8.5.1.2). A key measure for this work is, thus, the management of global and local power distributions. To this end, Knechtel and Sinanoglu develop efficient algorithms for floorplanning-centric voltage assignment, a technique well-known for power management in 2D ICs [Ma+11], but so far mostly overlooked for 3D ICs.

For any floorplan layout, the authors initially estimate the timing paths. They do so because the prospects for voltage assignment depend primarily on timing slacks—the more slack a module has, the lower the voltage one may apply, and the more controllable its power and thermal footprints become. The authors estimate the net delays via the well-known Elmore delays (here with consideration of wires and TSVs), and the delays of modules are estimated as proposed in [Lin10]. Note that more detailed timing analysis is not practical at the floorplanning stage, where placement and routing of gates are not available yet, but only high-level module properties.

Next, voltage volumes, i.e., the generalized 3D version of voltage domains spanning across multiple dies, are constructed. Knechtel and Sinanoglu do so by considering each module individually as the root for a multi-branch tree representation of voltage volumes. Each tree/volume is recursively build up via a breadth-first search across the respectively adjacent modules. During this merging procedure, the authors update the resulting set of feasible voltages, i.e., the voltages one may commonly apply for all modules within the volume without violating any timing/delay constraints. As result, a finalized tree represents a set of practical voltage volumes: Each node comprises a volume, and all the modules and voltages are encoded in the node and its ancestors. In order to limit the recursive process to relevant volumes and, hence, to save runtime and memory cost, the authors bound/prune parts of trees in two cases: (i) only the highest voltage would remain applicable for the volume after merging, which provides no power saving and typically incur large gradients; or (ii) merging the next module into the volume would (a) alter the set of feasible voltages and (b) induce the largest spatial overlap with the volume, among all adjacent modules to select from for merging. In other words, (ii) is to avoid large overlaps among different voltage domains, which would incur high power-domain routing cost [LMC09].

In short, the steps above are to determine all practical voltage volumes along with their modules, and also to track the timing and power values for the modules when assigned to those volumes. Finally, Knechtel and Sinanoglu select voltage volumes such that they can optimize for (i) locally uniform power densities within volumes, and (ii) small power gradients across volumes, as motivated by their initial findings (Sect. 8.5.1.2). To this end, the authors target for lowest standard deviations of power values both within voltage volumes and across volumes.

Knechtel and Sinanoglu found that their techniques induce a low runtime cost, around 30%, when compared to 3D floorplanning without voltage assignment. This is noteworthy because previous work [Lin10, LCL14] employ computationally expensive MILP formulations which render their integration into 3D floorplanning flows impractical in the first place.

Sampling of Activities and Post-processing   To impersonate an attacker triggering various activity patterns by alternating the inputs at runtime, Knechtel and Sinanoglu model the power profiles of all modules as Gaussian distributions. Without loss of generality, the authors set up each distribution with the module's nominal power value as mean and a standard deviation of 10%. The authors stepwise evaluate the steady-state temperatures using HotSpot [ZSS15] and sample the correlation stability (Eq. 8.5) in 100 runs over the entire 3D IC.

Continuing the runtime sampling process, the authors iteratively insert dummy thermal TSVs where the most stable correlations occur, as long as the resulting average correlation is reduced. This stop criterion represents the final "sweet spot" where further TSV insertion would increase the overall correlation again (Sect. 8.5.3.2).

8.5.3   Experimental Evaluation

8.5.3.1   Setup

Knechtel and Sinanoglu consider selected *GSRC* and *IBM-HB+* benchmarks, reviewed in Table 8.3. Note that the authors scale up the modules' footprints in order to obtain sufficiently large dies; 3D ICs are only promising to excel 2D ICs for large-scale integration [KL16, KYL15, Lim13]. The resulting die outlines are fixed, making the floorplanning problem practical yet challenging [KYL15]. For stacking of the two dies, the authors assume the face-to-back fashion [Lim13, ITRS16] (see also Fig. 8.10, or Fig. 6.1 in Chap. 6). Further technical details such as material properties and TSV dimensions are given in the respective default configurations of [ZSS15, Kne16].

Each benchmark is floorplanned 50 times in two different setups: (*i*) power-aware floorplanning and (*ii*) TSC-aware floorplanning. It is important to note that setup (*i*) provides a competitive and challenging baseline for the evaluation of (*ii*). Unlike the authors consider in (*i*), most previous work did not account for continuous optimization of the voltage assignment within the optimization loops, and may have thus missed on the full potential of power-aware 3D floorplanning. For (*i*), Knechtel and Sinanoglu optimize the packing density, wirelength, critical delay, peak temperature, and voltage assignment, all at the same time; all criteria are weighted equally. For voltage assignment, the authors seek to minimize both the overall power and the number of required voltage volumes. For (*ii*), the authors

Table 8.3 Considered *GSRC* and *IBM-HB+* benchmarks

Name	# Modules (hard/soft)	Modules' scale factor	# Nets	# Terminal pins	Outline [mm^2]	Power (for 1.0 V) [W]
n100	(0/100)	10	885	334	16	7.83
n200	(0/200)	10	1585	564	16	7.84
n300	(0/300)	10	1893	569	23.04	13.05
ibm01	(246/665)	2	5829	246	25	4.02
ibm03	(290/999)	2	10,279	283	64	19.78
ibm07	(291/829)	2	15,047	287	64	9.92

consider the same criteria as for (i), rendering it a practical and competitive setup. Additionally, the authors consider the thermal leakage and seek to minimize both the average correlation coefficients and the average spatial entropies. Here, for voltage assignment, the authors now seek to minimize (a) the number of required voltage volumes and (b) the standard deviations of power gradients among and across different volumes. For implementing the voltage volumes in both setups (i) and (ii), the authors consider three options: 0.8 V, with power scaling of $0.817\times$ and delay scaling of $1.56\times$; 1.0 V, without impact on power/delay; and 1.2 V, with $1.496\times$ power scaling and $0.83\times$ delay scaling. These values are simulated for the 90 nm node [Lin10].

8.5.3.2 On Destabilizing the Leakage Correlation

A low runtime stability of correlations is crucial to thwart TSC attacks (Sect. 8.5.2.2): The lower the correlation for various inputs, the less likely an attacker succeeds when modeling the thermal leakage. Knechtel and Sinanoglu observe that their techniques significantly reduce both the runtime stability of correlations and the nominal, steady-state correlations. Consider Fig. 8.13 as an example. In order to meet timing constraints, some modules (colored red

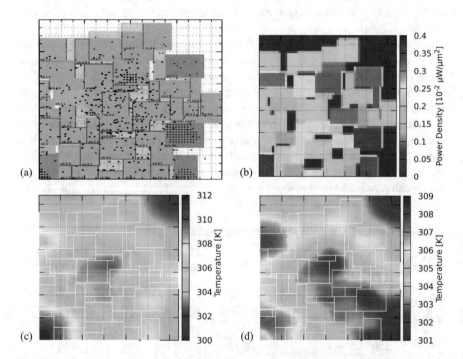

Fig. 8.13 (**a**) An exemplary floorplan for the bottom die of benchmark *n100*, (**b**) its power distribution, the thermal maps (**c**) before and (**d**) after TSV insertion by post-processing

in Fig. 8.13a) had to have high voltages assigned. The accordingly high power consumption of those modules disrupts the otherwise more uniform power distribution (Fig. 8.13b). As a result, for varying activity patterns triggered by an attacker, the highest stability of correlations occurs around those modules. The post-processing stage consequently inserts dummy TSVs there (see black dots in Fig. 8.13a). For the nominal case, i.e., for average activity patterns, the resulting shift in thermal behavior is reflected in Fig. 8.13c,d. In this example, the correlation coefficient drops from 0.461 to 0.324. This means that an attacker seeking to retrieve sensitive activities based on thermal readings is on average ≈30% less likely to succeed.

Knechtel and Sinanoglu note that post-processing should proceed until it reaches a "sweet spot" of lowest correlation. More specifically, the insertion of dummy thermal TSVs may stabilize at some point the correlation again, by inducing adverse side-effects on previously more decorrelated regions. For example, consider the upper left corner in Fig. 8.13c,d: The local correlation there increased to some degree, due to the insertion of TSVs in the other, previously highly correlated regions like the lower right corner. To account for such trade-off effects, recall that the post-processing stage inserts dummy TSVs only as long as the overall correlation decreases. The authors argue that, alternatively, one may adapt that stage to focus on reducing the correlation stability primarily for the critical module(s) to be protected from TSC attacks, and to accept more stable correlations elsewhere.

8.5.3.3   Leakage Trends and Mitigation Rates

To understand general trends on the thermal leakage in 3D ICs and the mitigation rates offered by their techniques, Knechtel and Sinanoglu study the average ranges of spatial entropies and correlation coefficients over 50 runs (Fig. 8.14 and Table 8.4). However, it is important to note that these numbers do not reflect well on best cases, which a designer will select carefully depending on the needs for security and the margin for design cost; see Sect. 8.5.3.4 for the latter.

When comparing the power-aware and the TSC-aware results, the key observation is that the latter exhibit lower correlations for the bottom die (r_1), by 16.79% for the largest *GSRC* benchmark *n300*, by 15.25% for the largest *IBM-HB+* benchmark *ibm03*, and by 7.71% over all benchmarks on average. While the average reduction over all benchmarks may appear somewhat low, its impact is important. It translates directly to 7.71% higher noise on average for an attacker, and to an accordingly higher degree of freedom for a designer to employ sensitive modules. Besides, a noteworthy trend here is the scalability: The larger the circuit, the lower the correlation achieved by the proposed TSC-aware floorplanning.

Both the power-aware and the TSC-aware setup can achieve low correlations for the bottom die, but at the cost of notably higher correlations for the top die. This holds also true in cases where the spatial entropies in the top die are much lower than those in the bottom die. This limitation is due to the following: Since the heatsink is attached above the top die, *Corblivar* [KYL15, Kne16] by default employs a

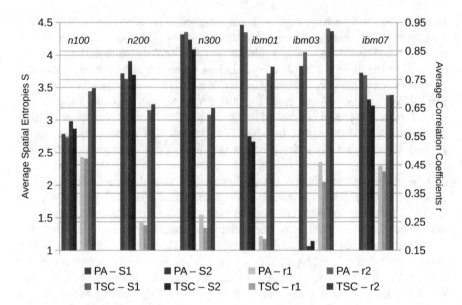

Fig. 8.14 Spatial entropies (S1, S2) and correlation coefficients (r1, r2) for power-aware floorplan-ning (PA) and thermal side-channel-aware floorplanning (TSC). Illustrated are the average ranges over 50 floorplanning runs; see also Table 8.4

thermal design rule to place high power modules preferably into the top die. In turn, this results in large power gradients across the two dies as well as within the top die. Large gradients in the top die naturally increase its correlation. Furthermore, the large disparities between the two dies obfuscate any "heat injection" from the bottom die into the top die which may otherwise decrease the top die's correlation, e.g., see Fig. 8.11e–h versus Fig. 8.11a–d. The authors find that, while one may relax *Corblivar*'s design rule, this can prohibitively increase the peak temperatures.

8.5.3.4   Design Impact and Effectiveness

Knechtel and Sinanoglu also study the average impact on other design criteria (Table 8.4). Again, these results are relevant to understand general trends, while they do not represent particular corner cases.

Recall that a key measure in the proposed methodology is the management of global and local power distributions, realized by tailored voltage assignment. The proposed technique increases power only by 5.38% on average when compared to power-aware floorplanning, whose key criterion is minimal power. Despite this (limited) increase of power, the authors observe notable reductions of the steady-state peak temperatures, namely by 13.22% on average (i.e., with respect to the ambient temperature of 293 K). The authors consider this as a beneficial side-effect of their proposed, multi-objective, and iterative 3D floorplanning flow. Besides, the

Table 8.4 Average spatial entropies, correlation coefficients, and average design cost over 50 runs of power-aware floorplanning (top) versus thermal side-channel-aware floorplanning (bottom)

Metric	GSRC benchmarks			IBM-HB+ benchmarks			
	$n100$	$n200$	$n300$	$ibm01$	$ibm03$	$ibm07$	Avg.
Power-aware floorplanning							
Spatial entropy S_1	2.787	3.718	4.315	4.462	3.830	3.727	3.806
Correlation coefficient r_1	0.476	0.249	0.274	0.199	0.459	0.446	0.351
Spatial entropy S_2	2.982	3.905	4.237	2.752	1.061	3.314	3.041
Correlation coefficient r_2	0.708	0.642	0.625	0.771	0.928	0.694	0.728
Overall power [W]	7.890	7.801	12.522	4.291	27.038	10.737	11.713
Critical delay [ns]	0.833	0.784	1.164	1.49	3.54	2.82	1.771
Wirelength [m]	30.671	29.086	37.100	21.491	57.999	108.020	47.394
Peak temp [K] [ZSS15]	309.811	308.475	311.095	303.496	334.759	308.715	312.725
Signal TSVs	451	897	1100	3489	3809	10,740	3414
Dummy thermal TSVs	–	–	–	–	–	–	–
Voltage volumes	6.900	12.365	14.365	3.675	5.333	3.027	7.610
Runtime [s]	83	218	262	489	197	507	226
Thermal side-channel-aware floorplanning							
Spatial entropy S_1	2.735	3.635	4.350	4.348	4.040	3.689	3.799
Correlation coefficient r_1	0.471	0.238	0.228	0.189	0.389	0.426	0.324
Spatial entropy S_2	2.868	3.698	4.086	2.668	1.140	3.219	2.946
Correlation coefficient r_2	0.719	0.662	0.649	0.795	0.919	0.695	0.739
Overall power [W]	10.331	8.172	12.967	4.520	26.829	11.247	12.344
Critical delay [ns]	0.878	0.817	1.162	1.607	3.831	3.434	1.954
Wirelength [m]	32.956	28.733	33.928	21.913	57.449	112.468	47.907
Peak temp [K] [ZSS15]	315.563	309.246	312.168	303.459	311.066	309.201	310.117
Signal TSVs	451	897	1099	3490	3806	10,741	3414
Dummy thermal TSVs	64	29	25	3	16	144	47
Voltage volumes	9.705	17.048	19.658	14.256	12.909	11.888	14.244
Runtime [s] 287	531	524	1075	374	667	576	

authors observe an average impact on other criteria as follows: Wires are 1.08% longer, critical delays increase by 10.33%, and, most notably, 87.17% more voltage volumes are implemented. Finally, Knechtel and Sinanoglu note that (i) power- and TSC-aware floorplanning exhibit the same average number of signal TSVs, and that (ii) TSC-aware floorplanning employs only few additional thermal TSVs (1.37% on average) during post-processing. Both findings indicate that the proposed TSC-aware floorplanning realizes the reported low correlation not only by carefully inserting dummy TSVs, but more so by thoroughly exploring the 3D design space.

8.6   Closing Remarks

Side-channel attacks are arguably one of the most severe threats to data and design confidentiality, as they can take advantage of the unintentional correlations and information leakage resulting from standard CMOS design practices. On the other hand, emerging technologies have proven to be inherently secure against these attacks due to their radically different working principles and operational dynamics. For instance, logic circuits constructed from SiNWFETs and ASL devices are able to mitigate PSC attacks owing to their uniform power profiles. Further, the low overheads of some emerging technologies allow designers to add extra logic for smoothening the power traces and eliminate any data dependencies. This chapter provided an overview of the basic concepts and considerations to achieve side-channel resilience with emerging technologies, and reviewed some examples of prior implementations in this field.

The case studies highlighted in this chapter investigated the resiliency of an emerging technology against power side-channel attacks, and addressed the leakage of secret activity patterns within 3D ICs. The former study showed that the NCFET technology can help to render some AES cipher hardware more resilient against the classical correlation power attack, and that the device-specific properties like the ferroelectric thickness play an important role in this resiliency. The second study modeled information leakage via thermal patterns as a built-in criteria into floorplanning and sought to minimize it, using novel and efficient design techniques and heuristics, thereby incorporating this notion of security early on in the 3D-IC design flow.

References

[Amr+18] H. Amrouch et al., Negative capacitance transistor to address the fundamental limitations in technology scaling: processor performance. IEEE Access **6**, 52754–52765 (2018). ISSN: 2169-3536. https://doi.org/10.1109/ACCESS.2018.2870916

[AYL18] Q. Alasad, J. Yuan, J. Lin, Resilient AES against side-channel attack using all-spin logic, in *Proceedings of the 2018 on Great Lakes Symposium on VLSI* (2018), pp. 57–62

[BCO04] E. Brier, C. Clavier, F. Olivier, Correlation power analysis with a leakage model, in *Proceedings of the Cryptographic Hardware and Embedded Systems* (2004)

[Ben+16] F. Beneventi et al., Thermal analysis and interpolation techniques for a logic + wideIO stacked DRAM test chip. Trans. Comput. Aided Des. Integr. Circuts Syst. **35**(4), 623–636 (2016). ISSN: 0278-0070. https://doi.org/10.1109/TCAD.2015.2474382

[Cla+16] L.T. Clark et al., ASAP7: a 7-nm FinFET predictive process design kit. Microelect. J. **53**, 105–115 (2016). https://doi.org/10.1016/j.mejo.2016.04.006

[Cla05] C. Claramunt, A spatial form of diversity, in *Proceedings of the International Conference on Spatial Information Theory*, Ellicottville (2005), pp. 218–231. https://doi.org/10.1007/11556114_14

[Cou+16] P. Coudrain et al., Experimental insights into thermal dissipation in TSV-based 3-D integrated circuits. Des. Test **33**(3), 21–36 (2016)

[Dem+12] J. Demme et al., Side-channel vulnerability factor: a metric for measuring information leakage. SIGARCH Comp. Arch. News **40**(3), 106–117 (2012). ISSN: 0163-5964. https://doi.org/10.1145/2366231.2337172

[Fei+15] Y. Fei et al., A statistics-based success rate model for DPA and CPA. J. Cryptogr. Eng. **5**, 227–243 (2015)

[Fu+17] Y. Fu et al., Kalman predictor-based proactive dynamic thermal management for 3D NoC systems with noisy thermal sensors. Trans. Comput. Aided Des. Integr. Circuts Syst. **99**, 1–1 (2017). ISSN: 0278-0070. https://doi.org/10.1109/TCAD.2017.2661808

[GG18] E. Giacomin, P.-E. Gaillardon, Differential power analysis mitigation technique using three-independent-gate field effect transistors, in *2018 IFIP/IEEE International Conference on Very Large Scale Integration (VLSI-SoC)* (IEEE, Piscataway, 2018), pp. 107–112

[HS14] M. Hutter, J.-M. Schmidt, The temperature side channel and heating fault attacks, in *Smart Card Research and Advanced Applications*, vol. 8419. Lecture Notes in Computer Science (Springer, Berlin, 2014), pp. 219–235. ISBN: 978-3-319-08302-5. https://doi.org/10.1007/978-3-319-08302-5_15. http://link.springer.com/chapter/10.1007%2F978-3-319-08302-5_15

[INK11] T. Iakymchuk, M. Nikodem, K. Kępa, Temperature-based covert channel in FPGA systems, in *Proceedings of the International Workshop on Reconfigurable Communication-Centric Systems-on-Chip* (2011), pp. 1–7. https://doi.org/10.1109/ReCoSoC.2011.5981510

[ITRS16] International Technology Roadmap for Semiconductor 2.0. ITRS (2016). http://www.semiconductors.org/main/2015_international_technology_roadmap_for_semiconductors_itrs/

[KDK14] G. Khedkar, C. Donahue, D. Kudithipudi, Towards leakage resiliency: memristor-based AES design for differential power attack mitigation, in *Machine Intelligence and Bio-inspired Computation: Theory and Applications VIII*, vol. 9119 (International Society for Optics and Photonics, Bellingham, 2014), p. 911907

[Kha+17] M.N.I. Khan et al., Side-channel attack on STTRAM based cache for cryptographic application, in *2017 IEEE International Conference on Computer Design (ICCD)* (IEEE, Piscataway, 2017), pp. 33–40

[KJJ99] P. Kocher, J. Jaffe, B. Jun, Differential power analysis, in *Advances in Cryptology* (1999), pp. 388–397

[KL16] J. Knechtel, J. Lienig, Physical design automation for 3D chip stacks – challenges and solutions, in *Proceedings of the 2016 on International Symposium on Physical Design* (2016), pp. 3–10. https://doi.org/10.1145/2872334.2872335

[Kne+20] J. Knechtel et al., Power side-channel attacks in negative capacitance transistor. IEEE Micro **40**(6), 74–84 (2020). https://doi.org/10.1109/MM.2020.3005883

[Kne16] J. Knechtel, Corblivar floorplanning suite and benchmarks (2016). https://github.com/IFTE-EDA/Corblivar

[Kne20] J. Knechtel, Correlation power attack. Extended from "Side Channel Analysis Library" by Y. Fei et al. Retrieved originally in March 2019 from https://tescase.coe.neu.edu/?current_page=SOURCE_CODE&software=aestool.2019-2020. https://github.com/DfX-NYUAD/CPA

[KPS19a] J. Knechtel, S. Patnaik, O. Sinanoglu, 3D integration: another dimension toward hardware security, in *Proceedings of the International Symposium on On-Line Testing and Robust System Design* (2019), pp. 147–150. https://doi.org/10.1109/IOLTS.2019.8854395

[Kri+17] Z. Krivokapic et al., 14 nm ferroelectric FinFET technology with steep subthreshold slope for ultra low power applications, in *Proceedings of the International Electron Devices Meeting* (2017), pp. 15.1.10–15.1.4

[KYL15] J. Knechtel, E.F.Y. Young, J. Lienig, Planning massive interconnects in 3-D chips. Trans. Comp. Aided Des. Integ. Circ. Sys. **34**(11), 1808–1821 (2015). https://doi.org/10.1109/TCAD.2015.2432141

[LCL14] B. Lee, E.-Y. Chung, H.-J. Lee, Voltage islanding technique for concurrent power and temperature optimization in 3D-stacked ICs, in *Proceedings of the International Technical Conference on Circuit/Systems Computers and Communications* (2014), pp. 267–269. http://dtl.yonsei.ac.kr/dtl_publications.html

[Lim13] S.K. Lim, *Design for High Performance, Low Power and Reliable 3D Integrated Circuits* (Springer, Berlin, 2013). https://doi.org/10.1007/978-1-4419-9542-1

[Lin10] H.-L. Lin, A multiple power domain floorplanning in 3D IC. M.A. Thesis. National Tsing Hua University (2010). http://handle.ncl.edu.tw/11296/ndltd/50629662251624775179

[LMC09] W.-P. Lee, D. Marculescu, Y.-W. Chang, Post-floorplanning power/ground ring synthesis for multiple-supply-voltage designs, in *Proceedings of the International Symposium Physical Design* (2009), pp. 5–12

[Ma+11] Q. Ma et al., MSV-driven floorplanning. Trans. Comput. Aided Des. Integr. Circuts Syst. **30**(8), 1152–1162 (2011). ISSN: 0278-0070. https://doi.org/10.1109/TCAD.2011.2131890

[Mas+15] R.J. Masti et al., Thermal covert channels on multi-core platforms, in *Proceedings of the USENIX Security Symposium* (2015), pp. 865–880. ISBN: 978-1-931971-232. https://www.usenix.org/conference/usenixsecurity15/technical-sessions/presentation/masti

[Mül+12] J. Müller et al., Ferroelectricity in simple binary ZrO2 and HfO2. Nano Lett. **12**(8), 4318–4323 (2012). PMID: 22812909. https://doi.org/10.1021/nl302049k

[OD19] C. O'Flynn, A. Dewar, On-device power analysis across hardware security domains. Trans. Cryptogr. Hardw. Embed. Sys. **2019**(4), 126–153 (2019). https://doi.org/10.13154/tches.v2019.i4.126153

[Pae+13] S. Paek et al., PowerField: a probabilistic approach for temperature-to-power conversion based on Markov Random field theory. Trans. Comp. Aided Des. Integ. Circ. Syst. **32**(10), 1509–1519 (2013). ISSN: 0278-0070. https://doi.org/10.1109/TCAD.2013.2272542

[Pah+16] G. Pahwa et al., Analysis and compact modeling of negative capacitance transistor with high ON-current and negative output differential resistance—part II: model validation. Trans. Electron. Dev. **63**(12), 4986–4992 (2016)

[Pat+18a] S. Patnaik et al., Advancing hardware security using polymorphic and stochastic spin-hall effect devices, in *Proceedings of the Design Automation Test in Europe*, 97–102 (2018). https://doi.org/10.23919/DATE.2018.8341986

[Pen+15] Y. Peng et al., Thermal impact study of block folding and face-to-face bonding in 3D IC, in *Proceedings of the International Interconnect Technology Conference* (2015), pp. 331–334

[Sam+16] S.K. Samal et al., Adaptive regression-based thermal modeling and optimization for monolithic 3-D ICs. Trans. Comput. Aided Des. Integr. Circuts Syst. **35**(10), 1707–1720 (2016). ISSN: 0278-0070. https://doi.org/10.1109/TCAD.2016.2523983

[SD08] S. Salahuddin, S. Datta, Use of negative capacitance to provide voltage amplification for low power nanoscale devices. Nano Lett. **8**(2), 405–410 (2008). https://doi.org/10.1021/nl071804g

[SW12] S. Skorobogatov, C. Woods, In the blink of an eye: there goes your AES key, in *IACR Cryptology ePrint Archive*, vol. 296 (2012)

[TS15] M. Taha, P. Schaumont, Key updating for leakage resiliency with application to AES modes of operation. IEEE Trans. Inf. Forensics Secur. **10**(3), 519–528 (2015). https://doi.org/10.1109/TIFS.2014.2383359

[Van+11] G. Van der Plas et al., Design issues and considerations for low-cost 3-D TSV IC technology. IEEE J. Solid State Circuits **46**(1), 293–307 (2011)

[XC17] Q. Xu, S. Chen, Fast thermal analysis for fixed-outline 3D floorplanning. Integ. VLSI J. **59**, 157–167 (2017)

[Xie+16] Y. Xie et al., Security and vulnerability implications of 3D ICs. Trans. Multi Scale Comp. Sys. **2**(2), 108–122 (2016)

[YSS13] J.J. Yang, D.B. Strukov, D.R. Stewart, Memristive devices for computing. Nat. Nanotechnol. **8**(1), 13–24 (2013)

[ZF05] Y.B. Zhou, D.G. Feng, Side-channel attacks: ten years after its publication and the impacts on cryptographic module security testing. IACR Cryptology ePrint Archive 388 (2005). http://eprint.iacr.org/2005/388

[Zhu+08] C. Zhu et al., Three-dimensional chip-multiprocessor run-time thermal management. Trans. Comput. Aided Des. Integr. Circuts Syst. **27**(8), 1479–1492 (2008). ISSN: 0278-0070. https://doi.org/10.1109/TCAD.2008.925793

[ZSS15] R. Zhang, M.R. Stan, K. Skadron, HotSpot 6.0: validation, acceleration and extension. Technical Report. University of Virginia (2015). http://lava.cs.virginia.edu/HotSpot/index.htm

[Fic+13] D. Fick et al., Centip3De: a cluster-based NTC architecture with 64 ARM cortex-M3 cores in 3D stacked 130 nm CMOS. IEEE J. Solid State Circuits **48**(1), 104–117 (2013). ISSN: 0018-9200. https://doi.org/10.1109/JSSC.2012.2222814

[Iye15] S.S. Iyer, Three-dimensional integration: an industry perspective. MRS Bull. **40**(3), 225–232 (2015). ISSN: 1938-1425. https://doi.org/10.1557/mrs.2015.32

[Kim+12] D.H. Kim et al., 3D-MAPS: 3D massively parallel processor with stacked memory, in *Proceedings of the International Solid-State Circuits Conference* (2012), pp. 188–190. https://doi.org/10.1109/ISSCC.2012.6176969

[Shi18] A. Shilov, AMD Previews EPYC Rome Processor: Up to 64 Zen 2 Cores (2018). https://www.anandtech.com/show/13561/amd-previews-epyc-rome-processor-up-to-64-zen-2-cores

[PSF17] V.F. Pavlidis, I. Savidis, E.G. Friedman, *Three-Dimensional Integrated Circuit Design*, 2nd edn. (Morgan Kaufmann Publishers Inc., Burlington, 2017). https://www.sciencedirect.com/book/9780124105010/three-dimensional-integrated-circuit-design

[P D+17] S.M. P D et al., A scalable network-on-chip microprocessor with 2.5D integrated memory and accelerator. Trans. Circ. Sys. I **64**(6), 1432–1443 (2017). ISSN: 1549-8328. https://doi.org/10.1109/TCSI.2016.2647322

[Kim+19] J. Kim et al., Architecture, chip, and package co-design flow for 2.5D IC design enabling heterogeneous IP reuse, in *Proceedings of the Design Automation Conference* (2019). https://doi.org/10.1145/3316781.3317775

[Sto+17] D. Stow et al., Cost-effective design of scalable high-performance systems using active and passive interposers, in *Proceedings of the International Conference on Computer-Aided Design (ICCAD)* (2017). https://doi.org/10.1109/ICCAD.2017.8203849

[Cle+16] F. Clermidy et al., New perspectives for multicore architectures using advanced technologies, in *Proceedings of the International Electron Devices Meeting (IEDM)* (2016), pp. 35.1.1–35.1.4. https://doi.org/10.1109/IEDM.2016.7838545

[Tak+13] S. Takaya et al., A 100 GB/s wide I/O with 4096b TSVs through an active silicon interposer with in-place waveform capturing, in *Proceedings of the International Solid-State Circuits Conference Digest of Technical Papers* (2013), pp. 434–435. https://doi.org/10.1109/ISSCC.2013.6487803

[Lau11] J.H. Lau, The most cost-effective integrator (TSV interposer) for 3D IC integration system-in-package (SiP), in *Proceedings of the ASME InterPACK* (2011), pp. 53–63. https://doi.org/10.1115/IPACK2011-52189

[Viv+20] P. Vivet et al., A 220GOPS 96-core processor with 6 chiplets 3D-stacked on an active interposer offering 0.6 ns/mm latency, 3Tb/s/mm2 inter-chiplet interconnects and 156 mW/mm2@ 82%-peak-efficiency DC–DC converters, in *2020 IEEE International Solid- State Circuits Conference* (2020), pp. 46–48. https://doi.org/10.1109/ISSCC19947.2020.9062927

[DY18] J. Dofe, Q. Yu, Exploiting PDN noise to thwart correlation power analysis attacks in 3D ICs, in *SLIP '18: Proceedings of the 20th System Level Interconnect Prediction Workshop* (2018). https://doi.org/10.1145/3225209.3225212

[KS17] J. Knechtel, O. Sinanoglu, On mitigation of side-channel attacks in 3D ICs: decorrelating thermal patterns from power and activity. Proceedings of the 54th Annual Design Automation Conference (2017), pp. 12:1–12:6. https://doi.org/10.1145/3061639.3062293

[Gu+16] P. Gu et al., Thermal-aware 3D Design for side-channel information leakage, in *Proceedings of the International Conference on Computer Design* (2016), pp. 520–527. https://doi.org/10.1109/ICCD.2016.7753336

Index

© The Author(s), under exclusive license to Springer Nature Switzerland AG 2021
N. Rangarajan et al., *The Next Era in Hardware Security*,
https://doi.org/10.1007/978-3-030-85792-9

Printed in the United States
by Baker & Taylor Publisher Services